Practical Amateur Astronomy
Celestial Objects for Modern Telescopes

Amateur astronomy is entering a new and exciting era, and this completely modern and up-to-date guidebook has been written for those star-gazers who wish to make the most of the latest technology.

Based on field notes made by the author during his own career as an amateur astronomer, this unique guide covers both traditional and novel approaches to studying the night sky. In addition to the more-standard techniques, it discusses the latest modern resources available to today's astronomer, such as personal computers, the Internet, and computerized telescopes. It includes practical advice on aspects such as site selection and weather; provides the reader with detailed instructions for observing the Sun, Moon, planets, and all types of deep-sky objects; and introduces newer specialties such as satellite observing and the use of astronomical databases. The book concludes with detailed information and observing tips for 200 interesting stars, clusters, nebulae, and galaxies, specially chosen to be visible with modest-sized telescopes under suburban conditions.

A new book for a new type of observing, *Celestial Objects for Modern Telescopes* will form the vanguard of new books in this area for the twenty-first century. Written to complement the author's other recent book *How to Use a Computerized Telescope*, this book will also appeal to astronomers with more-traditional equipment.

MICHAEL COVINGTON, an avid amateur astronomer since age 12, has degrees in linguistics from Cambridge and Yale. He does research on computer processing of human languages at The University of Georgia, where his work won first prize in the IBM Supercomputing Competition in 1990. His current research and consulting areas include theoretical linguistics, natural language processing, logic programming, and microcontrollers. Although a computational linguist by profession, he is recognized as one of America's leading amateur astronomers and is highly regarded in the field. He is the author of several books, including the highly acclaimed *Astrophotography for the Amateur* (1985; second edition, 1999) and *How to Use a Computerized Telescope* (2002), which are both published by Cambridge University Press. The author's other pursuits include amateur radio, electronics, computers, ancient languages and literatures, philosophy, theology, and church work. He lives in Athens, Georgia, USA, with his wife Melody and daughters Cathy and Sharon, and can be visited on the Web at www.covingtoninnovations.com.

Practical Amateur Astronomy

Celestial Objects for Modern Telescopes

Michael A. Covington

CAMBRIDGE
UNIVERSITY PRESS

PUBLISHED BY THE PRESS SYNDICATE OF THE UNIVERSITY OF CAMBRIDGE
The Pitt Building, Trumpington Street, Cambridge, United Kingdom

CAMBRIDGE UNIVERSITY PRESS The Edinburgh Building, Cambridge CB2 2RU, UK
40 West 20th Street, New York, NY 10011-4211, USA
477 Williamstown Road, Port Melbourne, VIC 3207, Australia
Ruiz de Alarcón 13, 28014 Madrid, Spain
Dock House, The Waterfront, Cape Town 8001, South Africa

http://www.cambridge.org

First published 2002

Printed in the United Kingdom at the University Press, Cambridge

Typefaces Palatino 10/13 pt and Meta Plus book   *System* LaTeX 2_ε [TB]

A catalogue record for this book is available from the British Library

ISBN 0 521 52419 9 paperback

Soli Deo gloria

Contents

Contents

Contents

Preface

Together with its companion volume, *How to Use a Computerized Telescope*, this book is a guide for a new generation of amateur astronomers. The two books began as a single project, originally a list of interesting objects that I put together for my own use at the telescope. Soon I added a concise summary of the Meade LX200 operating manual. Simon Mitton of Cambridge University Press saw my notes and encouraged me to turn them into a book. As the project grew, it became two books instead of one – a book about telescopes and a book about the sky. This volume is the latter.

While I was writing the two books, Scott Roberts of Meade Instruments lent me equipment to try out. The technical support departments at Meade, Celestron, Software Bisque, and Starry Night Software answered technical questions. Daniel Bisque supplied software for testing. Howard Lester, Dennis Persyk, Lenny Abbey, Rich Jakiel, T. Wesley Erickson, Robert Leyland, R. A. Greiner, Richard Seymour, Ralph Pass, Phil Chambers, Ells Dutton, Michael Forsyth, and John Barnes critiqued drafts of parts of the text. Tom Sanford let me try out his Meade LX90 at length. Earlier, Jim Dillard first got me interested in computer-aided astronomy by buying my old Meade LX3 from me and outfitting it with digital setting circles. There are probably others whose names I've forgotten to list, and I beg their indulgence. And I have hopelessly lost track of who helped with which volume! I also want to thank Janet Mattei of the AAVSO and Maurice Gavin of the B.A.A. for help with variable-star charts and spectrograms respectively. My wife Melody and my daughters Cathy and Sharon provided valuable encouragement and displayed admirable patience.

Please visit me on the Web at http://www.covingtoninnovations.com, where this book will have its own web page with updates and related information.

Michael Covington
Athens, Georgia
December 25, 2001

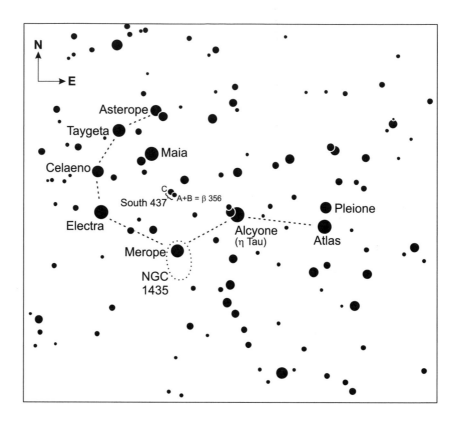

Part I
Amateur astronomy

Chapter 1
Using this book effectively

1.1 Amateur astronomy for a new generation

This is a handbook for the modern amateur astronomer. As far as possible, I've tried to write the book that I'd like to have in my own hands while at the telescope – along with a star atlas and the *Handbook of the B.A.A.,* of course.

Amateur astronomy isn't what it used to be. A generation ago, most serious amateurs observed from their homes with large Newtonians; one star atlas and two or three reference books were the amateur's complete guide to the sky; the latest news, arriving by magazine, was two months old; and most of the stars visible in the telescope were absent from even the largest catalogues and atlases.

Those days are gone, thank goodness. Telescopes have changed – they are nearly all portable, and compact designs such as the Schmidt–Cassegrain are popular. As often as not, the telescope is computer-controlled.

More importantly, computers have brought high-quality data sources within the amateur's reach. Alongside star atlases, we use software that plots the star positions measured by the Hipparcos satellite. We can compute the positions of comets, asteroids, and artificial satellites at the touch of a button. We can even track clouds by satellite to see if we're going to have clear weather.

Accordingly, a major theme of this book is the effective use of astronomical data, especially the Internet. Web addresses are given throughout, as well as detailed information about classic and modern catalogues of celestial objects. The book's web site, http://www.covingtoninnovations.com, will give ongoing updates.

1.2 The maps are backward!

On a more mundane level, the bulk of amateur telescopes now have star diagonals or flip mirrors, so that they present an image that is right-side-up but flipped left to right (Figure 1.1, bottom).

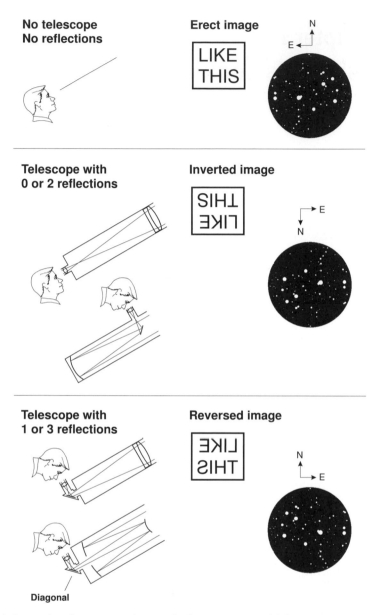

Figure 1.1. Image in telescope may be upside down or reversed left to right. Most of the maps in this book show the view through a telescope with one or three reflections (bottom).

Accordingly, my maps of the Moon (p. 25), Mars (p. 50), Jupiter (p. 53), and most telescopic star fields show the image with north up but flipped east to west, to match the view in the telescope. Users of Newtonians, and anyone who still uses a refractor without a diagonal, will have no trouble using numerous maps available elsewhere. On the art of matching up a flipped image with a non-flipped map, see p. 95.

1.3   Old books

Like many amateur astronomers, I enjoy comparing my observations to those of Admiral Smyth, T. W. Webb, Sir John Herschel, and other nineteenth-century observers who used telescopes about the same size as mine. To facilitate this, I have included information on obsolete constellation names (p. 86), precession of coordinates (p. 249), and old magnitude scales (p. 114), as well as information on classic observing guides, some of which are readily available in reprint (pp. 161–164).

1.4   Material you can skip

Smaller type, like this, indicates technical material that you can skip until you need it. By printing it in smaller type, I avoided having to take it out of its logical place in the text.

1.5   Pronouncing foreign names

One thing that frustrated me, when I was new to astronomy, was the vast number of foreign names I didn't know how to pronounce. Indeed, that may have been one of the things that led me to take up linguistics instead of astronomy as a career!

To keep you from being equally frustrated, in this book I indicate the pronunciation of every name likely to give difficulty. I even consulted native speakers of Danish and Czech, respectively, to placate years of curiosity about how to pronounce *Tycho Brahe* (p. 102) and *Antonín Bečvář* (p. 94).

Latin presents special problems. The ancient Roman pronunciation (with *v* pronounced as *w*, etc.) was reconstructed around 1900 but is rarely used by scientists. Arguably, it is not appropriate for names that originated after Roman times.

For familiar historical and mythological names, including those of the constellations, planets, and satellites, I follow the "English method." That is how Latin was pronounced in England in the 1800s, and it remains the traditional way to pronounce Latin names in English context.

For less familiar Latin names, such as lunar and planetary features, I use a "Continental" (late medieval) pronunciation that is similar to that of the Romans but pronounces *c, g,* and *v* as in English and *ae* and *oe* like Latin *e.* This pronunciation would have sounded natural to Copernicus and Kepler and is still widely used in Europe.

For full Latin pronunciation rules, ancient and modern, see William T. Stearn, *Botanical Latin* (David & Charles, 1995).

Though modern astronomers need not learn Latin, one piece of classical erudition that you still can't do without is the Greek alphabet (p. 91). Memorize it right away or be puzzled by every star chart.

Chapter 2
Observing sites and conditions

2.1 Darkness and night vision

2.1.1 Dark adaptation

The human eye takes time to adapt to dim light. Although the pupil opens up almost immediately, that's not the whole story. Dark adaptation involves the release of the light-sensitive chemical **rhodopsin** (visual purple) in the retina. For astronomy, useful dark adaptation takes about ten minutes, and substantial improvement continues for half an hour or more.

The central part of the retina does not function in dim light; faint objects disappear if you look straight at them. Experienced observers use **averted vision,** which means that they view the faintest stars, nebulae, and galaxies by looking slightly to one side of the object rather than directly at it.

Visual perception of faint objects is not continuous. A star near the limit of vision may be evident only a third of the time; it will seem to pop into and out of view. As long as you keep seeing it in the same place, you can be sure that it's real even though you can't see it continuously.

Even a small amount of bright light prevents complete dark adaptation; that's why a distant streetlight or even an illuminated doorbell button can be so annoying. Red light does not do this as much as other colors, which is why astronomers use red flashlights. The light must be *red* (or orange), not just *reddish*; what matters is the absence of blue wavelengths, not the red color.

Red photographic safelights are ideal, as are red light-emitting diodes, but anything that looks pink or purplish is inadequate, and red lights should not be too bright. Red filters for flashlights can be made from Rubylith, a masking material sold at art supply stores.

Night vision is impaired by too much bright sunlight during the day, especially during the hours immediately before observing. Sunglasses can be a wise investment. During brief trips indoors while observing, or to start dark-adapting before you go outside, you can wear red goggles.

2.1.2 Twilight and moonlight

The sky is not completely dark when the Sun is less than 18° below the horizon. This period of semidarkness is called **astronomical twilight** and its exact duration can be computed with software or looked up in the *Astronomical Almanac*. The following rough guide is good enough for most purposes:

- At latitude +30° (Florida, Texas), astronomical twilight lasts about $1\frac{1}{2}$ hours year-round.
- At latitude +40° (New York), astronomical twilight lasts $1\frac{1}{2}$ hours most of the year, but 2 hours from May to August.
- At latitude +50° (Vancouver and southern England), astronomical twilight lasts about 2 hours most of the year, $2\frac{1}{2}$ hours in May and August, and never ends in June or July because the Sun is never 18° below the horizon, even at midnight.

The 18° limit is not hard and fast. Good astronomical photographs have been taken when the Sun was 15° or even 12° below the horizon. Particularly at suburban sites, there is no point in waiting for perfect darkness that will never come.

Complete darkness also requires the Moon to be absent; see p. 24 for more about its movements.

2.1.3 Light pollution

Because of city lights, many of us have to go out into the country to see the stars. Even the most avid astronomers don't want cities to be dark; artificial light makes driving safer and prevents burglary and vandalism. But much of the light sent into the sky by city lights is completely unnecessary. This wasted energy is called **light pollution**. Far too much light is directed into people's eyes or into the sky rather than onto the objects that were meant to be illuminated.

Most non-astronomers don't care whether lights block the view of the sky, but there are other ways to motivate them to oppose light pollution. Most importantly, unshielded outdoor lights don't do their job. When you look down a street, you should not see streetlights – you should see the street! We have lampshades indoors; why not outdoors?

What's more, waste is waste. Light that goes up into the sky is wasted energy, and somebody is paying for it, both in money and in pollution from power plants. Properly shielded lighting provides better visibility at lower cost.

Campaigns against light pollution are conducted by the International Dark-Sky Association (3225 N. First Ave., Tucson, AZ 85719, U.S.A., http://www.darksky.org) and the British Astronomical Association (see p. 30). These organizations can also advise you about shields that can be added to streetlights to direct more of their light toward the ground.

The best way to deal with a bothersome streetlight near your site is to contact the local authorities and offer to pay for such a shield. Some amateur astronomers turn streetlights off by directing a laser pointer at the photocell, but this should only be done on private property with the full consent of everyone affected. In 2001, a Maryland amateur was fined $450 for tampering with county property when a neighbor innocently reported that a streetlight was out, and the repair crew saw his laser beam. If the absence of light had been blamed for a crime or accident, he might have faced a costly lawsuit.

Light pollution does not interfere with lunar and planetary observing, though there is some benefit from going out into the country to get away from hot pavement, rooftops, air conditioners, and the resulting unsteady air.

Nor does light pollution prevent you from using large telescopes. A big telescope is better than a small one at the same site, no matter where that site may be. Note however that one of the main uses of really large amateur telescopes – 16 inches (40 cm) and larger – is to view faint galaxies. That can't be done in the city because the surface brightness of the sky greatly exceeds that of the galaxies, regardless of the telescope.

2.1.4 Naked-eye limiting magnitude

The clarity of the air varies dramatically from night to night. Factor-of-two variations between seemingly clear nights often go unnoticed.

One way to measure transparency, as well as quality of the observing site, is to find the magnitude of the faintest stars that can be seen by dark-adapted eyes without a telescope. This is called the naked-eye limiting magnitude (NELM) and ranges from about 4 in the city to 5.5 at a good small-town site, 6.2 in the country, and 6.5 to 7 in the desert. (For explanation of the magnitude system, see p. 112.)

Figures 2.1–2.4 show the magnitudes of stars in selected regions of the sky. You can use them to determine your NELM. Deep-sky observers always want the NELM to be as faint as possible, but in practice, 5.0 to 5.5 is good enough for serious observing.

Remember that an *intermittently* visible star should be counted; a star at the limit of visibility will seem to flicker in and out of view. A more practiced observer will get a fainter NELM than a less experienced observer at the same site.

2.1.5 The Bortle dark-sky scale

Taking the idea of NELM further, John Bortle has defined a quality scale for rating dark-sky observing sites (*Sky & Telescope,* February, 2001, pp. 126–129). Table 2.1 on p. 11 sums it up. Remember that the top two classes are rarely seen, and most amateur observing is done in skies of Classes 4 to 6. Kitt Peak National Observatory is somewhere around Class 2 or 3.

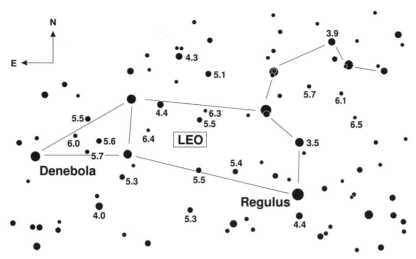

Figure 2.1. Naked-eye limiting-magnitude map of Leo, high in the spring evening sky. Adapted from a chart generated with *TheSky* software, copyright 2001 Software Bisque, Inc., used by permission.

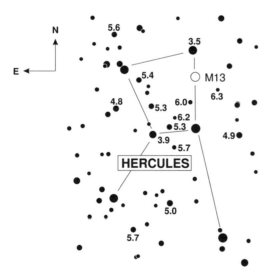

Figure 2.2. Naked-eye limiting-magnitude map of Hercules, high in the summer evening sky. Adapted from a chart generated with *TheSky* software, copyright 2001 Software Bisque, Inc., used by permission.

Notice that there is a sharp transition around NELM 5.0. Going from 5.0 to 5.5 is a big improvement; going from 5.5 to 6.0 or even 6.5 is not so dramatic. The reason is that 5.0 is the approximate limiting magnitude of an experienced observer *without dark adaptation,* or at least without much dark adaptation. Thus, 5th-magnitude stars are visible, or almost visible, even under poor conditions, but it takes a much darker sky to get to 5.5 or 6.

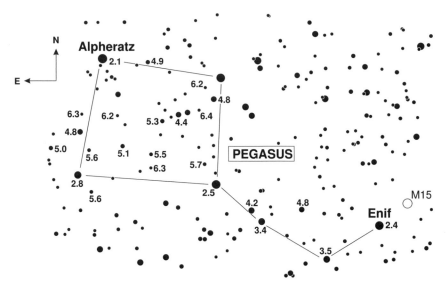

Figure 2.3. Naked-eye limiting-magnitude map of Pegasus, high in the autumn evening sky. Adapted from a chart generated with *TheSky* software, copyright 2001 Software Bisque, Inc., used by permission.

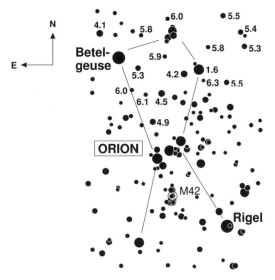

Figure 2.4. Naked-eye limiting-magnitude map of Orion, high in the winter evening sky. Adapted from a chart generated with *TheSky* software, copyright 2001 Software Bisque, Inc., used by permission.

Many of Bortle's tests involve the visibility of the **zodiacal light,** a cloud of interplanetary dust that looks a bit like twilight, extending upward along the ecliptic (the plane of the solar system) after true twilight is over. The zodiacal light is about half as bright as the Milky Way, and under extremely dark, clear skies, extends to form a **zodiacal band** all along the ecliptic, with a bright spot,

Table 2.1. *The Bortle dark-sky scale*

Class	NELM	Description
1 (Excellent)	>7.6	Rarely seen. Milky Way spectacular; zodiacal band (along ecliptic) plainly visible; no light on ground (cars, telescopes invisible).
2 (Truly dark)	7.1–7.5	Remote western deserts. Milky Way elaborately structured; zodiacal light prominent along ecliptic in the west after end of twilight; M33 and many other Messier objects seen without telescope.
3 (Rural)	6.6–7.0	Some light pollution on horizon, none overhead. Milky Way prominent; zodiacal light easily seen. Good country skies in the eastern U.S. and Britain.
4 (Rural–suburban)	6.1–6.5	Definite domes of light pollution on horizon. Milky Way clearly visible; brightest part (in summer) shows considerable structure. Telescopes and cars are visible from across the field.
5 (Suburban)	5.6–6.0	Milky Way visible but only brightest parts are prominent. Obvious sources of light pollution in several directions. Still good enough for serious observing and astrophotography.
6 (Bright suburban)	≈5.5	Milky Way somewhat hard to see. M31, M44 visible to naked eye but not prominent. Clouds, when present, are bright, illuminated from below. All Messier objects are visible in a 5-inch (12.5-cm) telescope, but more serious deep-sky observing and photography are not feasible.
7 (Suburban–urban)	≈5.0	Entire sky grayish, not black. Milky Way invisible or very hard to see. M31, M44 barely visible without a telescope. Deep-sky enthusiasts should concentrate on multiple stars, clusters, and planetary nebulae.
8 (City)	≈4.5	Gray or orangish sky. Objects on the ground are very clearly visible by reflected skylight. Many constellations unrecognizable because so many stars are hidden.
9 (Inner city)	≤4.0	Only the brightest stars are visible. Planets, Orion, and Ursa Major may still be picked out. Fainter objects can only be found with the help of a computerized telescope.

Table 2.2. *Scale of atmospheric steadiness devised by E. M. Antoniadi (1870–1944)*

I	Perfect seeing, without a quiver
II	Slight undulations, with moments of calm lasting several seconds
III	Moderate seeing, with larger air tremors
IV	Poor seeing, with constant troublesome undulations
V	Very bad seeing, scarcely allowing the making of a rough sketch

the **Gegenschein** (*GAY-ghen-shine,* German for "counterglow") directly opposite the Sun. The zodiacal light is easiest to see after sunset in the spring and before dawn in the autumn, when the ecliptic is high in the sky.

2.2 Atmospheric steadiness

The air is much steadier at some times than at others. Lunar and planetary observers often rate steadiness ("seeing") on the Antoniadi scale (Table 2.2). In practice, Classes III and IV are the most common. My personal scale of "excellent," "good," "mediocre," and "poor" corresponds roughly to Classes II to V.

To a remarkable extent, steadiness depends on conditions close to the telescope and even inside it.

When a telescope is first taken outside, steadiness is almost always poor. (The first few minutes may be good, and then conditions deteriorate rapidly.) The view steadies again as the telescope reaches the temperature of the surrounding air. This typically takes an hour per 10 °C (18 °F) of temperature change – somewhat less for small telescopes, more for larger ones. When the air temperature is changing rapidly, the telescope may never reach equilibrium. Remember, too, that the dew cap may either improve or worsen the seeing depending on its temperature.

The surroundings of the telescope also affect steadiness, especially the last two meters (six feet) or so. The best seeing occurs on an island in a lake or when looking over a cliff, so that there is nothing nearby from which heat waves can rise. Hot pavement, roofs, chimneys, and air conditioner exhausts make the air very unsteady. My telescope pier is at the end of a paved driveway, and I improved the seeing substantially by placing a plastic picnic table just south of the pier, to block the hot air rising from the concrete.

Excellent steadiness occurs some of the time almost anywhere. I have only once experienced perfect seeing (Antoniadi I) – with a 6-inch (15-cm) reflector, on a roof at Valdosta State University where conditions ought to have been dreadful.

The amplitude of atmospheric turbulence can be measured with CCD cameras. It ranges from 5 arc-seconds (very bad) through 2″ (good), 1″ (excellent), and occasionally 0.5″ or better (practically perfect).

The experienced planetary observer learns to look for moments of clarity – "revelation peeps" as Percival Lowell called them – transient though they may be. Some CCD cameras can do the same thing by taking pictures repeatedly and saving only the best. In poor conditions, smaller telescopes have an advantage because they do not require as much air to be steady at once. Steadiness is always better high in the sky because you are not looking through as much air as when an object is near the horizon.

2.3 Weather and the astronomer

2.3.1 Climate, weather, and seasons

Observatory sites are chosen for the maximum number of clear nights. Elsewhere, conditions can be every bit as good, though not as much of the time. Most of the deep-sky objects that we observe were discovered by Sir William Herschel in England; many of the best amateur photographs of the planets are taken by Donald Parker in marshy Miami; and Jerry Lodriguss photographs faint deep-sky objects from New Jersey.

Table 2.3 summarizes some effects of weather on viewing conditions. Good transparency and good steadiness seldom occur together. The air is clearest right after the passage of a cold front, preferably a rainy one. The first of a series of clear nights is generally the clearest and the least steady.

Table 2.3. *Effect of weather on astronomy*

Good transparency and/or sky darkness
Recent passage of a cold front
Recent rain followed by clearing
Good daytime visibility (of mountains, trees, etc.)
Low humidity
Cold weather
Winter and spring (in eastern United States)
Early morning rather than evening sky (in summer)
Clouds over nearby cities but not over the observing site

Good steadiness
Monotonous weather (little change for several days)
Jet stream at least 500 km (300 miles) away
Stationary high pressure area
Daytime and nighttime temperatures close together
High humidity, fog, heavy dew
Haze, smog, and thin cirrus clouds
Indoor temperature $\approx$ outdoor temperature (for quick equalization)
Late summer and autumn (in eastern United States)

One predictor of good steadiness is that the daytime and nighttime temperatures are close together. Some of the steadiest air occurs when the day has been cloudy (hence cool) and the clouds clear suddenly at sunset. Apart from this, the air is usually unsteady during changing weather.

Do not judge transparency by the number of bright stars in the sky. On winter evenings we look out at the Orion arm of our galaxy and see a host of bright stars and star clusters, even under mediocre conditions. In the late spring and early summer, the sky overhead is outside the plane of our galaxy, and the only bright stars are Regulus, Arcturus, and Spica. You may think the sky is hazy when it isn't, simply from the lack of bright stars. But it is true that the sky is often clearest in the winter and can be quite hazy in the summer.

Steadiness also varies with the seasons. One factor is of course telescope cool-down. In the summer, when indoor and outdoor temperatures are nearly the same, the telescope does not take long to equalize with the surrounding air, and steadiness seems to be better than in winter, when a long cool-down is required. In reality, the winter air can be quite steady once the telescope is properly adjusted to it.

The haze of summer – annoying as it is for other purposes – often brings very steady air. Some of my best double-star observing has been done under haze or cirrus clouds so heavy that only the brightest stars were visible to the naked eye. Under such conditions, a computerized telescope is a great help. The haze often diminishes after midnight, giving a reasonably transparent sky by dawn.

One of the most unnerving sights of summer, until you get used to it, is reflected lightning in a clear night sky. In the southeastern United States, "sheet lightning" or "heat lightning" (as some people call it) is rather common. The flashes are reflected from thunderstorms at a distance of 40 to 80 miles (60 to 120 km). Reflected lightning has surprisingly little effect on observing or even astrophotography; the storms are much too far away to be dangerous to people or equipment. But if you can hear thunder, a storm is nearby and you should pack up and go home.

2.3.2 Using satellite weather data

Will tonight be clear? Amateur astronomers quickly learn not to trust weather forecasts. The surest way to check weather conditions is to look at weather satellite data online at http://www.ghcc.msfc.nasa.gov/GOES/, http://www.weather.com, and other sites.

Two kinds of data are available. Figure 2.5 shows a visible-light image with a resolution of 1 km, showing clouds exactly as they look from outer space. Side-lit cumulus clouds look like mountains. Images of this type are available only in the daytime.

Figure 2.6 shows the more familiar infrared satellite image. This is actually a temperature map and is available 24 hours a day. Clouds show up as white because they block the emission of heat from the Earth's surface and lower

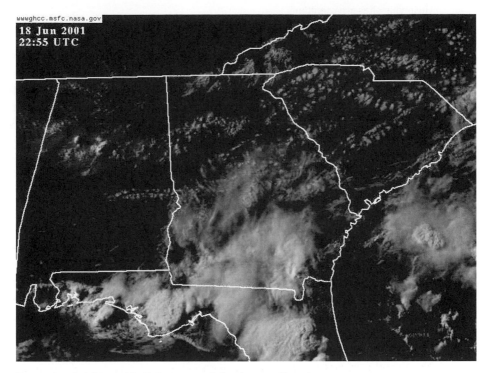

Figure 2.5. NASA visible-light image of clouds over Georgia.

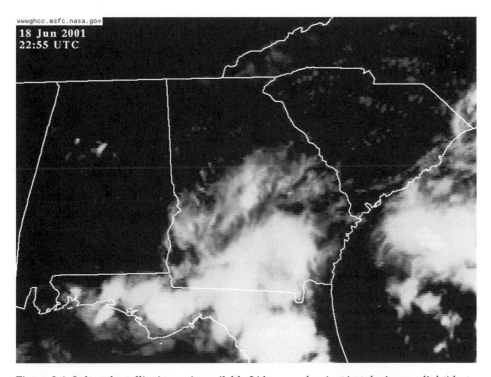

Figure 2.6. Infrared satellite image is available 24 hours a day (not just during sunlight) but has lower resolution and is more likely to mistake humid air for clouds.

15

atmosphere. Although magnified to the same size here, the infrared map has lower resolution.

Infrared maps tend to pick up high haze and regions of high humidity even when the air is still almost transparent. Visible-light maps, on the other hand, tend to miss high, thin clouds. The best way to learn to interpret both kinds of satellite maps is to compare them regularly to the real sky at your location.

2.3.3 Dew

The **dew point** is the temperature at which water begins to condense out of the air. Clearly, when the air cools below the dew point, dew will fall.

Unfortunately, that's not the whole story. Telescopes and other objects can easily get colder than the surrounding air because they cool by radiation. In the daytime, solid objects acquire heat from the Sun; at night, they re-radiate that heat back into space. Through radiation, a telescope can get cool enough to collect dew even when the air temperature is as much as 6 °C (10 °F) above the dew point. The corrector plate of a Schmidt–Cassegrain is particularly vulnerable. Dew on the mirror of a Newtonian is rare.

There are several remedies. One is to warm the afflicted lens with electrical heating elements. Not only does this remove dew, it can improve steadiness by bringing the glass back up to the temperature of the air around it. Two or three watts of power are usually sufficient. You can build your own dew heater out of a string of eight 10-ohm resistors in series, supplied with 12 volts. Commercial dew heaters are also available; one of the best is the Kendrick Dew Remover System (Kendrick Astro Instruments, 2920 Dundas St. West, Toronto, Ontario, Canada M6P 1Y8; http://www.kendrick-ai.com).

Another way to prevent dew is to shield the telescope lens from the sky by surrounding it with a cylindrical **dew cap** (Figures 2.7, 2.8), which should ideally be made of a material that does not conduct heat. I have used cardboard and plastic foam. You can make the dew cap more effective by adding a black paper lining that can be changed when it gets damp. In a pinch, the black paper itself can be the dew cap, secured by a rubber band or elastic cord. I have even made a dew cap out of a brown paper bag.

Dew is insidious; a light layer of it can degrade the view for a long time before you realize what's wrong. Once it has formed, you can get rid of it with warm air from a hair dryer, but doing so is likely to destroy thermal equilibrium – important for lunar and planetary observing but not for deep-sky work. Twelve-volt hair dryers or ice melters can be used in the field, but they deplete batteries quickly.

Even after dew has formed, simply installing the dew cap can make dew evaporate, especially if it has a dry, black paper lining. So can aiming the telescope at the ground. But dew caps do not ward off dew forever; they just delay it. Heating elements actually prevent dew from forming.

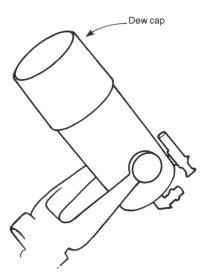

Figure 2.7. Dew cap can be made of black paper, foam, or any convenient material. (From *Astrophotography for the Amateur.*)

Figure 2.8. The dew cap of Cyrano de Bergerac, or perhaps Pinocchio. Half this length would be sufficient. This is plastic craft foam, but black paper would work almost as well.

Towels are useful during dewy observing sessions. Draping a dry towel across a lens (not touching it) can remove dew without changing the temperature. A dry beach towel can be thrown over the telescope when it is to be left unattended for a few minutes; another can be spread out on the table under the charts, notebooks, and eyepiece box.

Never put a lenscap on a dewy lens, and never store an equipment case in which dew has collected. You're likely to be greeted by rust and fungus the next time you open it. Instead, let everything air out indoors before closing it up. But if a lens is very cold and dew-free, put the cap on it before taking it indoors so that moisture will not condense on it there.

2.4 Observing at remote sites

2.4.1 Finding a site

By far the best way to find a dark country observing site is to join a local astronomy club or one in a nearby city. The Atlanta Astronomy Club, for instance, maintains several observing sites, and the Los Angeles Astronomical Society's site at Lockwood Valley is internationally famous.

A second strategy is to check nearby state and national parks, especially those that provide camping. Let the rangers know what you're doing, and in particular that you need complete darkness and don't need electricity. Some negotiation may be needed, and it helps to enlist other amateur astronomers. Across the United States and in other countries, amateur astronomers are pushing for state parks to recognize astronomy as a popular outdoor activity, alongside camping and fishing. Right now – during the computerized telescope boom – is the ideal time.

If you have a friend who owns a big ranch, then of course you don't have these problems. But observing on farms doesn't work as well as you might think. It's surprisingly hard to get completely away from 24-hour security lights.

2.4.2 Transporting the telescope

Telescopes are awkward to carry around. For a while, you can transport the telescope in the foam-lined box in which it came from the factory; a coat of paint on the outside will make the box last longer. Eventually, though, the box will wear out, and then you're on your own.

A case for the telescope may not be necessary. To transport a large Schmidt–Cassegrain in a small car, just put it in a seat as if it were a passenger, then throw a blanket over it, and fasten the seat belt.

For air travel or shipment by common carrier, however, a hard case is essential. Be forewarned that some of the foam-lined cases sold by telescope manufacturers are not actually big enough to hold the telescope with adequate padding. For example, my NexStar 5 does not fit into the case that came with it (which may have been intended for a smaller telescope). In any event, many telescope cases are actually camera cases available at much lower cost through camera stores. The foam in them is precut into partially detached squares which are easy to remove to make a cutout of any shape.

Good cases for smaller telescopes, with tough die-cut foam, are available from JMI (810 Quail Street, Unit E, Lakewood, CO 80215, U.S.A., http://www.jimsmobile.com). Larger cases of high quality – often costing $300 or more, but worth it – are available from companies that make custom cases for electronic equipment. One of many is Atlas Case Corporation, 1380 South Cherokee Street, Denver, CO 80223, http://www.atlascases.com; similar companies exist in most large cities. Cases of this type surround the telescope with several inches of foam in all directions and protect it from even the most ruthless baggage handlers.

There are cheaper solutions. When shopping for a custom case, ask if there are any slightly flawed or surplus cases of suitable size. (Unlike most purchasers, you don't need a matched set.) Also check the surplus and secondhand market. I was fortunate enough to get an extremely rugged military surplus shipping case that fits the foam in which my Meade LX200 was shipped.

Celestron telescopes are shipped in a different type of foam that is not meant to be reused and will eventually crumble, but you can buy new foam from a case manufacturer or from McMaster-Carr, 6100 Fulton Industrial Blvd., Atlanta, GA 30336, U.S.A., http://www.mcmaster.com; what you want is called "Unifoam" polyether-based polyurethane foam.

Naturally, you will also need an eyepiece case. Small foam-filled cases are available at camera stores and also (for much lower prices) at hardware stores. You can also buy replacement foam for existing cases.

Any case that is all black will get lost in the dark. I've added white reflective tape to all my equipment cases to make them more visible.

2.4.3 Site etiquette

The most important point of etiquette at observing sessions is to keep unwanted lights off. Use only red flashlights, and even then, do not aim them at people (nor at CCD cameras, which are very sensitive to red).

Car headlights are a special problem. Even fog lights and backup lights can disturb observers. I once drove a car out of Lockwood Valley, California (slowly!), using a handheld red flashlight for illumination, and braking with only the parking brake in order to keep the brake lights off.

In the US and Canada many newer cars have headlights that automatically come on when the car is in motion. Disabling them can be a challenge. A few vehicles have cutoff switches. More often, you have to remove a fuse or two. One trick that works with many cars is to depress the parking brake pedal slightly – just one or two clicks, not enough to engage the brake – before starting the engine. In desperate cases you may have to tape opaque plastic or paper on the front of the headlights. Try all of these things in advance, and if in doubt, park well away from the observers. Avoid braking and reversing as much as possible so that at least the brake lights and backup lights will stay off.

If you install switches in your car to disable any exterior lights, make sure you switch the lights on again when you get out on the road. It's a good idea to install red warning lights on the dashboard that come on when the exterior lights are disabled. Otherwise you could find yourself in a difficult situation if you were in an accident and someone alleged that you were driving without legally required lights.

Astronomers at observing sessions are usually quite busy; some of them will gladly show you things through their telescopes, but others are carrying out observing programs of their own, or even guiding photographs, and can't be disturbed. In this respect a star party is not a party in the usual sense of the word. It's a group work session.

Many astronomy groups object to smoking on the observing field for two reasons: it disturbs people, who are not at liberty to move away once their equipment is set up, and it's a fire hazard on dry grass. Loud noises or music are also likely to be unwelcome.

2.4.4 Keeping warm

It's *cold* at night. Here are my tactics for keeping warm:

1. Cover the whole body, not just the torso. I wear an insulated coverall from a hunting and fishing store, not a jacket. A warm cap is obligatory, too – and it must not have a brim, which would bump the telescope.
2. Dress in layers. Air is an excellent insulator. Leather is not; neither is thin, tight-fitting underwear. Long underwear should be loose and should preferably consist of more than one layer. Remember, you're dressing to stand still, not to ski; your best guidance will come from hunters and fishermen.
3. Don't get cold before putting on warm clothing. I climb into my coverall when the temperature falls below about 55 °F (12 °C). It's much easier to prevent a chill than to recover from one.
4. Don't get hungry or thirsty either – keep your metabolism up. Have a snack every three or four hours. Take brisk walks periodically.

2.4.5 Mosquitoes

Many observing sites are plagued by mosquitoes, particularly when the temperature remains above 20 °C (68 °F) all night. Mosquitoes are most active at dusk; most of them retire by midnight. They prefer shrubbery, tall grass, and marshy ground rather than dry, open spaces, and they seldom fly higher than about 2 meters (7 feet), so an elevated observing platform can avoid them altogether.

No site can be made permanently mosquito-free because mosquitoes fly for miles, but at least their local habitats can be eliminated. They will breed in anything that collects water, such as clogged gutters or old tires or paint cans, and they complete their whole life cycle in one week. At my home, the mosquito problem diminished greatly when I cut down a hollow tree that was about 2 meters from the telescope pier; mosquitoes were hiding in the leaves and breeding in the standing water inside the tree.

Mosquitoes are attracted by carbon dioxide (from human breath) and bite most readily on dry skin. Any skin moisturizer discourages them to some extent, as do garlic, thiamine (vitamin B1), spicy scents, and light-colored clothing. To really repel them, though, you need either oil of citronella, which has a strong smell and dissipates fairly rapidly, or DEET (N, N-diethyl-meta-toluamide), which is odorless and long-lasting but attacks paint and plastic. Some mosquitoes are insensitive to DEET; try several repellents until you find one that works on the local population. Ultrasonic devices are not effective; electric "bug zappers" may be worse than useless, since they destroy larger insects that prey on mosquitoes. The breeze from an electric fan, however, is quite effective.

Permethrin, an insecticide, both repels and kills mosquitoes; it is the active ingredient in Yard Guard and similar products but is cheaper as a liquid concentrate and is quite effective when sprayed around the observing site and onto nearby shrubbery. Spray a fine mist upward so that it reaches the underside of the leaves. Granular permethrin, sold as an ant killer, is good for keeping mosquitoes out of the grass directly underfoot. (And ants! Always check for anthills before setting up.) Naturally, nothing should be sprayed while optics are exposed.

Any doubts about the safety of insect repellents and insecticides should be balanced against the dangers of mosquito bites themselves; mosquitoes carry West Nile virus, encephalitis, and other diseases. DEET is apparently safe to use on skin when used as directed. Permethrin is not recommended for use on skin but has been used to impregnate army uniforms. A 3% solution of permethrin is used on the human body to kill lice; what I spray into the trees and shrubs is only 0.03%, and it's definitely strong enough to get rid of the mosquitoes.

2.4.6 Other vermin

There is always a chance of encountering other animals at a remote observing site, ranging from snakes to raccoons, skunks, and even (in the open-range West) cattle and wild horses. The best protection is to make your presence obvious: if alone, talk to yourself or play music. Most animals are eager to stay out of your way if not provoked. Strange smells, such as insect repellents and mothballs, also repel larger animals. Near the U.S.–Mexican border there is even some risk of encountering smugglers; they, too, are eager to stay out of your way if they can detect you from a distance.

2.4.7 Safety

I'd rather be an old coward than a short-lived daredevil, and my personal opinion is that no one should go to a remote site alone without a means of communication. Cellular telephones are cheap; ham and CB radio are cheaper. Also, someone should know exactly where you are and should be prepared to send help if you do not return on time. Periodic check-ins by radio or telephone are a good idea. Make sure that your communication system actually works; notoriously, cellphones don't work at remote sites.

The most likely misfortune is that your car won't start – which is a good reason for not powering your telescope from the car battery. Worse things can happen. In 1999, an American amateur, observing alone, tripped on ice, hit his head on the tailgate of his truck, and almost bled to death. Accidents of this type are rare, but one is too many.

Driving home after staying up all night can also be hazardous. Lack of sleep affects reaction time the same way as alcohol, and coffee is not a complete antidote; don't push your limits. My own practice – perhaps overcautious or perhaps just lazy – is to pack up and head home at 1 a.m. Other people camp on the observing site, sleeping all day while wearing dark masks.

Chapter 3
The Moon, the Sun, and eclipses

3.1 The Moon

3.1.1 Phases of the Moon

We often make jokes about activities that depend on the phases of the Moon, but amateur astronomy really does. When the Moon is high in the sky, especially if it is full or nearly full, you can't see faint stars, nebulae, or galaxies. Conversely, if you want to observe the Moon you need to know when and where it is going to appear.

Accordingly, all astronomical observers need to keep track of the phases of the moon. Figure 3.1 summarizes the whole cycle. The Casio "Forester" wristwatch, marketed to hikers and fishermen, keeps track of the cycle for you.

The Moon stays close to the ecliptic, though not precisely on it, and moves eastward, making a full circle relative to the stars every 27.32 days (one **sidereal month**). This means that it moves its own apparent width (half a degree) in slightly less than an hour. You can watch this happen when the Moon passes near a bright star.

The cycle of phases, or **synodic month**, takes 29.53 days. This is longer than the sidereal month because the Sun and Moon are moving in the same direction; the Moon has to move more than a full circle in order to catch up.

3.1.2 Why observe the Moon?

The Moon is a dead world, with no atmosphere, no active volcanoes, and far less seismic activity than the Earth. It has been studied with large telescopes, mapped by spacecraft, and even walked on by human beings. Surely there is nothing new for an amateur astronomer to see on the Moon – or is there?

That's how I felt for many years, but eventually I came to see three reasons for viewing the Moon with my telescope.

Age (Days past new moon)		Phase	Visibility
0	●	New	Not visible
1–6		Waxing crescent	In western sky after sunset; sets before midnight
7		First quarter	High in sky at sunset; sets at midnight
8–14		Waxing gibbous	Highest in late evening; sets before dawn
15	○	Full	Rises at sunset, highest at midnight, sets at sunrise
16–21		Waning gibbous	Rises after sunset, high in sky after midnight
22		Last quarter	Rises at midnight, high in sky at sunrise
23–28		Waning crescent	Rises after midnight; in eastern sky before sunrise
29	●	New	Not visible

Figure 3.1. The phases of the Moon.

Because it's there

First and most obviously, the Moon is a spectacular sight, impressive in even the smallest telescopes. If you like mountain scenery, you can enjoy the lunar mountains even from the plains of southern Georgia or the fens of Cambridgeshire.

Particularly striking features are the Alpine Valley, which cuts through the lunar Alps (Montes Alpes); the Straight Wall, which is actually a cliff; and the subtly undulating floor of the Bay of Rainbows (Sinus Iridum).

Despite being a satellite of Earth, the Moon is the size and shape of a small planet. No other planet-like body, not even Earth, shows us all its landforms at once. Because there is no erosion (except for the slow accumulation of meteorite impacts), the lunar surface is geologically fresh, inviting the observer to figure out how the features were formed. Its secrets are laid bare, if only we can decode them.

Because we never see the same view twice

If you think the Moon looks the same all the time, you haven't looked closely enough. The angle from which the Sun illuminates the Moon is constantly changing, so is the angle from which we view the lunar surface. You may never see the same exact combination twice.

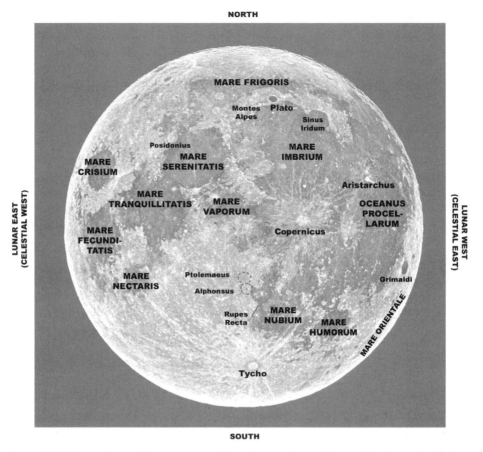

Figure 3.2. Mirror-image Moon map, matching the view in a telescope that has a diagonal. (Users of Newtonians can use ordinary Moon maps published elsewhere.)

Some changes are obvious. The **terminator**, the line separating sunlight from shadow, takes two weeks to move across the Moon's visible face, but by looking closely at mountain peaks and the insides of craters, you can see visible movement in just a couple of hours.

More detail is visible near the terminator than anywhere else. Many craters turn into white spots, or disappear from view entirely, when lit from directly above, at full moon. (Note that Ptolemaeus and Alphonsus had to be drawn in by hand on Figure 3.2.) Accordingly, your observing program must follow the phases of the Moon. View Mare Crisium when the Moon is a thin waxing crescent, Ptolemaeus and Alphonsus at first quarter, Copernicus a few days after first quarter, and Grimaldi just before full moon.

Although the Moon always keeps the same face toward the Earth, there is a slight wobble, called **libration**, because the Moon's orbit is not perfectly circular. Accordingly, features near the **limb** (the edge of the Moon's visible face), such as Mare Orientale, are much more visible at some times than at others. Once you begin observing the Moon regularly, you will notice that even the distance from Mare Crisium to the limb is quite variable.

Mare Orientale was originally classified as two mountain ranges, Montes Cordillera and Montes Rook. By observing at times of most favorable libration, H. P. Wilkins and Patrick Moore discovered in 1938 that the mountains curve around a plain, surrounded by concentric ramparts.

Space probes later revealed a spectacular system of three concentric circles around a giant impact basin similar to Mare Crisium. It turns out, in fact, that all lunar mountain ranges are the ramparts of impact sites, unlike Earth's mountain ranges, which are formed by movement of underground plates.

Because there are still unsolved mysteries

Transient lunar phenomena (TLPs, also called lunar transient phenomena, LTPs) are unexpected glows, discolorations, or mists obscuring lunar features. Their causes and even their existence are hotly disputed.

In 1958, Nikolai Kozyrev obtained a spectrogram of what appeared to be a volcanic glow on the central peak in the crater Alphonsus. In subsequent years the British Astronomical Association catalogued numerous TLPs, mostly involving the craters Alphonsus and Aristarchus, many of them confirmed by more than one observer.

We now know from space probes that there are no active volcanoes on the Moon, and that most of the craters were formed by meteorite impacts. (They are round because a powerful impact creates a circular shock wave regardless of whether the object hits head-on.) Apart from small meteorite impacts, the most that could be happening on the Moon today is that small puffs of radon or other gases are occasionally released, stirring up dust, when rocks are shifted by minor seismic activity or extremes of temperature.

In *Sky & Telescope,* September, 1999, William Sheehan and Thomas Dobbins argue that all TLPs are illusions or misinterpretations. They contend that Kozyrev's spectrogram did not actually show anything unusual. Rapid changes in brightness can occur during sunrise and sunset over the lunar surface, as mountain peaks are suddenly flooded with sunlight or hidden in shadow. Discolorations and mists are attributable to atmospheric effects on Earth, especially atmospheric dispersion, which can spread out any bright spot (such as Aristarchus) into a smear of color that is blue on one end and red on the other. One end of the smear is often hidden in the glare of other features, making the other end look like an isolated spot of color.

That, however, is not the whole story. In 1969, Apollo astronauts saw an unusual glow on the northwest wall of Aristarchus, and in 1971, Apollo 15 detected radon gas there. So one possibility is that occasional puffs of gas from inside the Moon raise thin clouds of dust.

Another possibility is that some parts of the lunar surface may reflect light anomalously at particular angles of illumination (compare with Martian flares, p. 51). We know that the Moon as a whole tends to reflect light in the direction from which it came, like a reflective-coated license plate, so that the full moon is much more than twice as bright as the half (quarter) moon. Perhaps there are patches of shiny surface material that show up as a glow when the sun hits them just the right way. If this is true, then TLPs are still of considerable scientific interest.

3.1.3 Names of lunar features

Figure 3.2 on p. 25 shows a mirror-image map of the Moon, to help you get oriented when using a telescope with a diagonal. Ordinary Moon maps have south at the top but are not mirror-imaged. Most of them show far more detail,

Figure 3.3. The Straight Wall (Rupes Recta, lower left of center) and the craters Ptolemaeus, Alphonsus, and Arzachel (above center). Taken with a 35-mm camera and 100-mm telephoto lens coupled afocally to an 8-inch (20-cm) $f/10$ Schmidt–Cassegrain with an 18-mm eyepiece. North is up; this is not a mirror image.

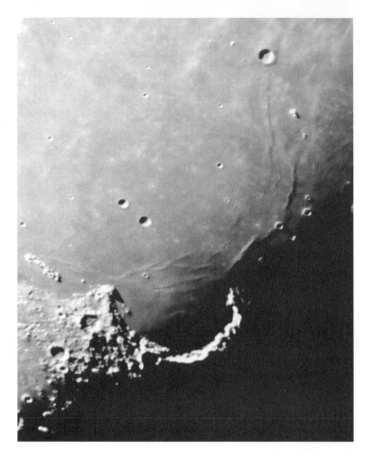

Figure 3.4. The Bay of Rainbows (Sinus Iridum) with its wavy floor. Same technique as Figure 3.3. South up, not mirror-imaged.

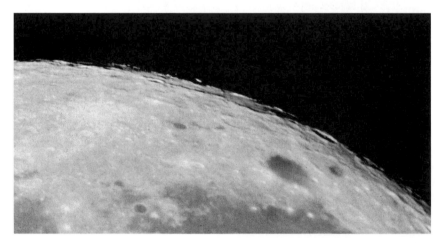

Figure 3.5. The concentric ramparts of Mare Orientale are visible in this picture taken under moderately favorable libration conditions. Better views are sometimes possible. Lunar southwest up; not mirror-imaged.

of course; the only purpose of this map is to prepare you to use a larger one. The definitive lunar atlas is *Atlas of the Moon*, by Antonín Rükl (Kalmbach, 1996; reprinted several times by different publishers). For an entertaining guide to the whole Moon, with maps and descriptions of features, I recommend *Patrick Moore on the Moon*, by (who else?) Sir Patrick Moore (published by Cassell, 2001), a revision of Moore's earlier *Guide to the Moon*.

Early telescopic observers thought the flat areas of the lunar surface were water, so they christened them *maria* (MAH-ree-ah, Latin for "seas"), and the name stuck. Mountain ranges are mostly named after mountains on Earth. Craters are named after individuals, usually scientists. The official nomenclature is managed by the International Astronomical Union, and additions are made periodically.

All the features except craters have Latin names; below are the pronunciations and meanings of the most important ones. These are "Continental" pronunciations (p. 5); alternatives are possible, and what is important is to accent the right syllable. Names of craters (e.g., Plato, Copernicus, Tycho, Herschel) are generally not Latin; they are pronounced like the names of the scientists whom they honor.

Mare	MAH-reh	
Anguis	AHN-gwiss	Serpent Sea
Cognitum	COG-nit-um	Known Sea (where Ranger 7 landed)
Crisium	CREE-see-um	Sea of Crises
Fecunditatis	fay-cun-dit-AH-tiss	Sea of Fertility
Frigoris	FREE-gor-iss	Sea of Cold
Humboldtianum	hum-bolt-ee-AHN-um	Humboldt's Sea
Humorum	hoo-MORE-um	Sea of Moisture
Imbrium	IM-bree-um	Sea of Rains
Insularum	in-soo-LAH-rum	Sea of Islands
Marginis	MAR-jin-iss	Border Sea
Nectaris	NECK-tar-iss	Sea of Nectar
Nubium	NOO-bee-um	Sea of Clouds
Orientale	oh-ree-ehn-TAH-leh	Eastern Sea
Serenitatis	seh-ren-it-AH-tiss	Sea of Serenity
Smythii	SMYTH-ee-ee	Smyth's Sea
Tranquillitatis	trahn-quill-it-AH-tiss	Sea of Tranquility
Undarum	un-DAH-rum	Sea of Waves
Vaporum	vah-PO-rum	Sea of Vapors
Montes Alpes	MON-tace AHL-pace	Alps Mountains
Oceanus Procellarum	oh-SEH-a-nus pro-sell-AH-rum	Ocean of Storms
Rupes Recta	ROO-pace RECK-ta	Straight Scarp ("Straight Wall")
Sinus	SEE-nus	
Iridum	EAR-id-um	Bay of Rainbows
Medii	MAY-dee-ee	Central Bay
Vallis Alpes	VAH-lis AHL-pace	Alpine Valley (in Montes Alpes)

Fecunditatis is also correctly spelled *Foecunditatis*, but *Undarum* is misspelled *Undarim* on one widely reproduced map.

3.1.4 Coordinate systems

As it revolves around the Earth, the Moon also rotates so that the same side of it faces us at all times. Its rotation defines its axis, with north and south poles, and its equator.

As Figure 3.2 indicates, two rival definitions of east and west have been used on the Moon. East in our sky is west as seen by an astronaut standing on the Moon's surface, and vice versa. The official designations were revised in 1961 so that Mare Orientale, the Eastern Sea, is now in the west.

Lunar (**selenographic**) latitude (β) and longitude (λ) are measured from the Moon's equator and prime meridian. The average center of the visible face is latitude $0°$, longitude $0°$. Latitude is positive to the north and negative to the south. Longitude is positive toward lunar east (Mare Crisium) and negative toward lunar west (or, alternatively, wraps around to $360°$).

Libration (more precisely, **optical libration**) is measured by the selenographic latitude and longitude (β', λ') of the center of the visible face of the Moon. It can exceed $6°$ in each direction. Parallax, caused by the fact that different places on Earth see the Moon from different angles, can add another degree; this is called **diurnal libration**. There is a small amount of **physical libration** caused by unevenness in the Moon's rotation.

There are two ways to measure the phase of the Moon. The **phase angle** is the angle between Earth, Moon, and Sun. The **Sun's selenographic colongitude** is the longitude of the sunrise terminator on the Moon, measured *westward* from $0°$ to $360°$ (thus $0°$ at first quarter, $90°$ at full moon, $180°$ at last quarter, and so on). Colongitude is tied to lunar features, which move around with libration, but phase angle is not; thus the two numbers do not measure exactly the same thing.

Colongitude, libration, and other physical parameters are given in the *Handbook of the British Astronomical Association* and the *Astronomical Almanac*, as well as software programs such as *TheSky*.

3.1.5 Observing programs

Along with many other amateur observing programs, lunar observing is co-ordinated by the British Astronomical Association (B.A.A., Burlington House, Piccadilly, London W1V 9AG, England, http://www.britastro.org) and the Association of Lunar and Planetary Observers (A.L.P.O., c/o Julius L. Benton, Jr., 305 Surrey Road, Savannah, GA 31410, U.S.A., http://www.lpl.arizona.edu/alpo).

Besides the TLP hunt, there are several research programs that involve studying lunar features under varying angles of illumination. A large area near the southeastern edge of the visible face was never adequately mapped by Apollo or Lunar Orbiter missions; it is known as Luna Incognita ("the unknown Moon") and visual observations are still needed.

Three good handbooks are Michael T. Kitt, *The Moon: An Observing Guide for Backyard Telescopes* (Kalmbach, 1992); Gerald North, *Observing the Moon* (Cambridge, 2000); and Peter T. Wlasuk, *Observing the Moon* (Springer, 2000).

The most basic observing technique is to draw the lunar features with pencil and paper, since you can always see more than you can photograph. The most common mistake is to choose too small a scale. A drawing should show a *tiny* area of the Moon, greatly magnified; if you draw a crater as large as Copernicus, it should fill half the page. (A good scale is 1 km on the Moon to 1 mm on the paper, two and a half times the size of the pictures in Rükl's atlas; your drawings are then exactly a millionth of actual size.) Outline drawings that show the positions of features are almost as useful as fully shaded renderings. Because of changing libration, they will often look quite different from photographs or maps.

3.1.6 Lunar eclipses

An eclipse of the Moon (Figure 3.6) takes place when the Moon passes into the Earth's shadow. If the Moon's orbit coincided with the ecliptic, there would be a lunar eclipse at every full moon, but because of the angle between the orbits and their constantly shifting orientation, lunar eclipses occur only a couple of times a year, and the Moon does not always go completely into the umbra (Figure 3.7).

A lunar eclipse is visible from anywhere on Earth where the Moon is in the sky at the time – which, in turn, means anywhere the Sun is not in the sky, since the Sun and Moon are directly opposite each other. Unlike solar eclipses, lunar eclipses do not require any eye protection.

Table 3.1 lists upcoming **umbral** eclipses, i.e., those in which the Moon passes through at least part of the umbra (the darkest part of the shadow). **Penumbral** eclipses are equally common but often go unnoticed because the darkening is so slight.

(a) As usually shown in textbooks:

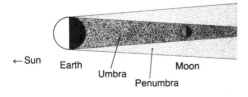

←Sun   Earth   Umbra   Penumbra   Moon

(b) A more accurate picture:

Figure 3.6. A lunar eclipse. (From *Astrophotography for the Amateur.*)

31

Table 3.1. *Umbral lunar eclipses, 2003–2010. Times are for umbral first and last contact; penumbral phases start earlier and end later. Dates as well as times are UT; in America, the local time generally falls on the previous evening. There are no umbral eclipses in 2002*

Date (UT)	Partial or total?	First contact (UT)	Last contact (UT)	Visible in USA?	Visible in UK?
2003 May 16	Total	02:03	05:18	Yes	Yes
2003 Nov. 8–9	Total	23:32	03:05	Yes	Yes
2005 Oct. 17	7%	11:34	12:32	Partly	No
2006 Sep. 7	19%	18:05	19:37	No	No
2007 Mar. 3	Total	21:30	01:12	Partly	Yes
2007 Aug. 28	Total	08:51	12:24	Partly	No
2008 Feb. 21	Total	01:43	05:09	Yes	Yes
2008 Aug. 16	81%	19:35	22:44	No	Yes
2009 Dec. 31	8%	18:52	19:53	No	Yes
2010 Jun. 26	54%	10:16	13:00	Partly	No
2010 Dec. 21	Total	06:32	10:02	Yes	Partly

Data from Fred Espenak, *Fifty Year Canon of Lunar Eclipses: 1986–2035* (NASA Reference Publication 1216).

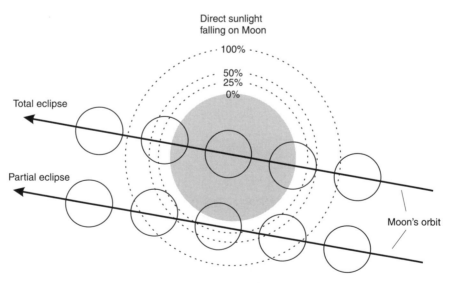

Figure 3.7. How a lunar eclipse looks from Earth. Moon darkens gradually in penumbra, then abruptly in umbra (shaded). (From *Astrophotography for the Amateur.*)

The appearance of the Moon during total eclipse depends on how much light is scattered by the Earth's atmosphere; the umbra is usually copper-colored, sometimes very dark, and occasionally lopsided. Its appearance is worth recording. A total lunar eclipse also gives you an opportunity to photograph the Moon surrounded by stars, which is impossible at any other time.

3.1.7 Occultations

An **occultation** is the passage of the Moon in front of a star, or occasionally a planet. Predictions are published in *Sky & Telescope*, in the *Handbook of the British Astronomical Association*, and online at http://www.lunar-occultations.com.

Stars always disappear at the (celestial) eastern edge of the Moon (away from Mare Crisium; the dark edge when the Moon is in the evening sky) and reappear less than an hour later at the western edge. Stars disappear instantly; planets disappear gradually over a few seconds; and double stars can disappear stepwise, first dimming as one component is covered, then vanishing completely a fraction of a second later. (Some double stars have been discovered this way.) Occasionally, when occulted by a jagged mountain range, a star will disappear and reappear repeatedly.

Occultations provide an excellent opportunity for precise measurement of positions in the Solar System, and observing is coordinated by the International Occultation Timing Association (c/o Craig and Terri McManus, 2760 S.W. Jewell Ave., Topeka KS, 66611-1614, U.S.A., http://www.occultations.org). Timings must be accurate to at least one second; to make them semi-automatically, use a video camera coupled afocally to the telescope, with sound fed in from a time signal station such as WWV (on shortwave at 5.0, 10.0, and 15.0 MHz).

3.2 The Sun

3.2.1 Sun filters

To view the Sun safely through a telescope, you must dim its light by a factor of about one million. This is equivalent to a logarithmic density of 6.0 (because $\log 1\,000\,000 = 6$). Lower densities, such as 5 or even 4, are safe when used as directed; they give a brighter image at higher powers. The important thing about the filter is not so much its visual density as the fact that it does not transmit excessive amounts of infrared or ultraviolet. For that reason, *only* filters designed for solar viewing should be used. Other filters, no matter how dark, generally transmit too much infrared.

The filter *must* be located in front of the telescope. Small sun filters for eyepieces were popular 30 years ago, but they were apt to overheat and crack, seriously injuring the observer. Also, the concentrated light of the sun can easily damage the inside of the telescope.

Good-quality sun filters are made of glass or plastic coated with a thin layer of metal. Filters of this type are now available from many vendors, including

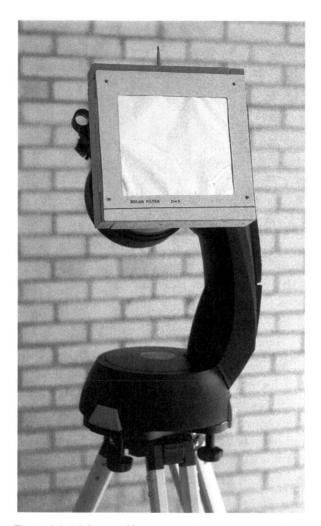

Figure 3.8. Mylar sun filter mounted in an improvised wooden cell.

Thousand Oaks Optical (Box 4813, Thousand Oaks, CA 91359, U.S.A., http://www.thousandoaksoptical.com) and Baader Planetarium (distributed by Astro-Physics, 11250 Forest Hills Road, Rockford, IL 61115, U.S.A., http:// www.astro-physics.com).

The cheapest way to get a good sun filter is to buy unmounted sheets of aluminized Mylar filter material for just a few dollars and make your own cell to hold the filter in front of the telescope. Mine, shown in Figure 3.8, is a bit over-elaborate; much simpler cardboard arrangements work well. The plastic should not be stretched tight; for best image quality, leave a bit of slack. In unsteady daytime air, many telescopes work better if masked down to about 3 or 4 inches (7 to 10 cm) aperture.

The filter *must not* fall off the telescope during use; make sure it can be mounted securely, and inspect it for holes every time you use it. Tiny pinholes,

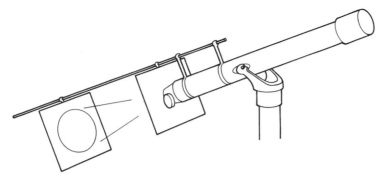

Figure 3.9. Sun projection should be attempted only with small refractors, and only when recommended by telescope manufacturer. Remove or cover finder to prevent damage.

though not dangerous, spoil the image by adding scattered light; they should be patched with spots of black paint.

Ordinary filters enable you to see sunspots, faculae, and granulation. To view solar flares, prominences, and elaborate surface detail, you need a narrow-band hydrogen-alpha (Hα) filter, available from Thousand Oaks (address just given) and from Coronado Filters (9121 East Tanque Verde Rd., Tucson, AZ 85749, U.S.A. http://www.coronadofilters.com). Narrow-band filters cost hundreds or thousands of dollars.

Another way to view the Sun is to project an image behind the eyepiece of a telescope (Figure 3.9). This technique should be used *only* with small re-fractors, *only* with simple or inexpensive eyepieces (no thick cemented lenses), and *only* when the manufacturer of the telescope recommends it. I have seen a Schmidt–Cassegrain whose secondary mirror came partly unglued from its support because of heating from sun projection; other kinds of reflectors can also be damaged. Even with a refractor, the finder must be covered to keep from burning up the crosshairs, and the telescope must be kept aimed straight at the Sun, so that concentrated sunlight never falls on the side of the tube.

3.2.2 Solar features

Figure 3.10 shows the main features of the Sun. The **photosphere** is the bright surface that we view through sun filters. The **chromosphere** (with its **promi-nences**) and the **corona** are much fainter outer layers that are visible only at total eclipses or with special equipment. The corona has no real boundary; it fades off into darkness several degrees away from the Sun.

Sunspots are dark areas perhaps half as bright as the rest of the photosphere; viewed in isolation, they would still look extremely bright. A typical sunspot consists of an **umbra**, in the center, and a **penumbra** with a pattern of radial streaks. Sunspots appear in groups, and even a single sunspot may turn out, when viewed with higher magnification, to be a tight group broken up by **light bridges** (bands of fully luminescent photospheric material).

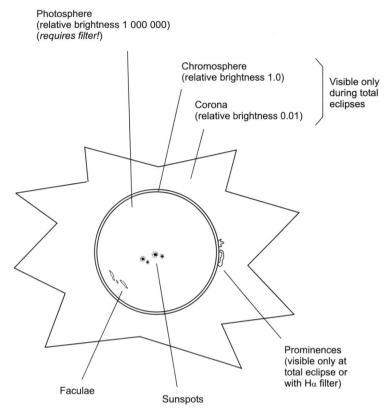

Figure 3.10. Anatomy of the Sun.

Sunspots come and go in an 11-year cycle. Maximum was reached in 2000, and the next minimum will occur around 2007, followed by a rise more rapid than the fall. Some sunspot groups are large enough to see with the unaided eye (protected by a filter, of course). Each spot lasts a few days or weeks. The first spots of a new cycle appear at high latitudes; as the cycle progresses, sunspot development moves toward the solar equator. Sometimes the first spots of a new cycle and the last spots of the old cycle can be seen at the same time.

The Sun rotates about every 25 days at the equator, but more slowly at higher latitudes, so that features initially aligned north–south will drift relative to one another. The average rotation period is 27.2753 days relative to the Earth. Rotation is from celestial east to west.

At maximum, the Sun outputs about 0.2% more energy than at minimum, just enough to cause minor shifts in the Earth's climate. (The Sun is thus a variable star, though a very low-amplitude one; the exact amount of its variability is disputed.) There was a prolonged lack of sunspots from 1645 to 1715, roughly coinciding with unusually cool weather in Europe.

At maximum the Sun also emits more charged particles, which are absorbed by the Earth's ionosphere, making it more reflective to radio waves; that's why international shortwave reception is better when sunspots are numerous. These particles also fuel the aurora borealis (Northern Lights).

Faculae are bright areas seen mainly near the limb, where limb darkening helps make them visible. Under good conditions you can also see **granulation** of the photosphere, sometimes hard to distinguish from film grain on photographs.

Under rare conditions, a white-light **solar flare** can be seen with an ordinary sun filter; it is an eruption within a sunspot group, visible as a bright spot that lasts a few minutes. Far more common are flares visible only in Hα light.

Solar observers are hampered by the fact that the daytime air is turbulent. Tactics to combat turbulence include observing over water or from the top of a cliff; observing near sunrise or sunset, when the air is steadier; and masking the telescope down to 4 inches (10 cm) or smaller.

Amateur observation of the Sun is coordinated by the American Association of Variable Star Observers (p. 132) and the British Astronomical Association.

3.2.3 Solar eclipses

A solar eclipse occurs when the Moon blocks the view of the Sun from some point on Earth. Directly under the Moon, the eclipse is **total** or **annular**, depending on whether the Moon is close enough to the Earth to block the whole Sun (Figure 3.11). For thousands of miles to either side, the eclipse is **partial**. Thus, a typical observing site will have partial eclipses every couple of years, but a total eclipse only once in several centuries.

Figure 3.12 shows what the observer sees. As long as any part of the photosphere is visible, a sun filter is necessary. As totality approaches, the last thin sliver of photosphere breaks up into **Baily's beads**, a series of disconnected spots caused by the Moon's jagged limb, and **shadow bands** caused by atmospheric turbulence are seen dancing on the ground. (There is some controversy, but it

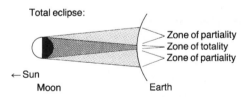

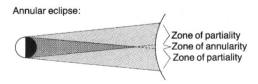

Figure 3.11. Solar eclipse configurations in space. Depending on the distance from the Moon to the Earth, the eclipse is total or annular. (From *Astrophotography for the Amateur.*)

Total eclipse

Annular eclipse

Any eclipse seen from outside the central path

Figure 3.12. How solar eclipses look from Earth. Any eclipse is partial if you are outside the path of totality or annularity. (From *Astrophotography for the Amateur.*)

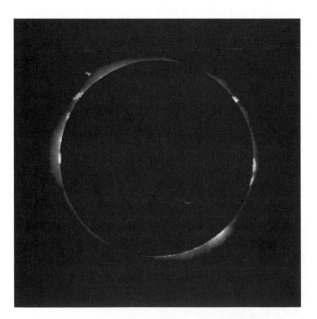

Figure 3.13. Solar prominences (one of them detached) and the inner corona are visible in this short exposure of the 1999 total eclipse in Turkey. (Kevin Chapman.)

Total Solar Eclipses: 2001 - 2020

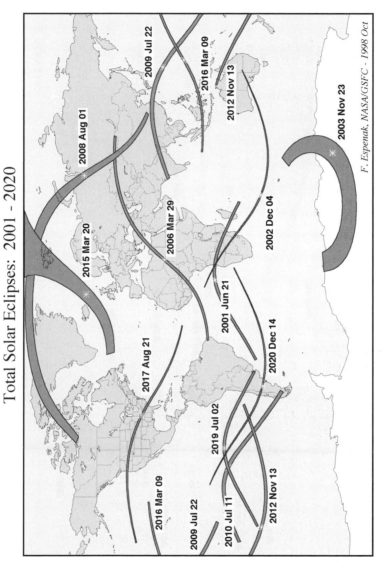

Figure 3.14. Total solar eclipses, 2001–2020. Predictions by Fred Espenak, NASA, from http://sunearth.gsfc.nasa.gov. Reproduced by permission.

Annular Solar Eclipses: 2001 - 2020

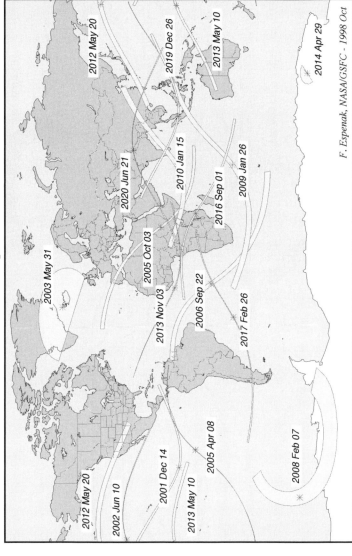

2012 May 20

2019 Dec 26

2013 May 10

2020 Jun 21

2010 Jan 15

2016 Sep 01

2009 Jan 26

2014 Apr 29

2003 May 31

2005 Oct 03

2013 Nov 03

2006 Sep 22

2017 Feb 26

2012 May 20

2002 Jun 10

2001 Dec 14

2005 Apr 08

2013 May 10

2008 Feb 07

F. Espenak, NASA/GSFC - 1998 Oct

Figure 3.15. Annular solar eclipses, 2001–2C20. Predictions by Fred Espenak, NASA, from
http://sunearth.gsfc.nasa.gov. Reproduced by permission.

appears that shadow bands are like pinhole-camera images of irregularities in the air, formed when the Sun is reduced to a narrow slit.) Then the chromosphere and corona come into view and the Sun can be viewed without filters. All too soon the eclipse is over; totality can never last much more than seven minutes, and one- and two-minute eclipses are more common.

Annular eclipses are less spectacular but well worth observing. Melody Covington was able to photograph part of the chromosphere at the 1984 annular eclipse in Georgia, and shadow bands put on quite a show.

Figures 3.14 and 3.15 show the paths of all solar and annular eclipses from 2001 to 2020. The continental United States does not get another total eclipse until 2017, but avid eclipse chasers will doubtless flock to Africa in 2006. Each of these eclipses is partial over a wide area, continent-sized or larger; details are published in the *Astronomical Almanac* and the *Handbook of the British Astronomical Association.* For copious eclipse information online, see Fred Espenak's site, http://www.mreclipse.com, and NASA's eclipse prediction site, http://sunearth.gsfc.nasa.gov.

Chapter 4
The planets

4.1 General concepts

The first thing a telescopic observer of the planets notices is that they all look rather small and blurry. Some training of the eye is required in order to see planetary detail. Not only are planetary features faint, but they require constant attention in order to take advantage of brief moments of steady air.

Drawing the planets is one of the best ways to learn to observe them (Figure 4.1). You can always see, and therefore draw, more detail than you can photograph with the same telescope. A good scale is about 5 cm (2 inches) for the diameter of the planet. Always draw what you see, not what you think you *ought* to see.

Every planetary drawing should be labeled with the date and time, particulars of the telescope, and eyepiece, and quality of seeing (atmospheric steadiness, Table 2.2, p. 12). Colored filters are often helpful. Such drawings have considerable scientific value and are collected for research purposes by the B.A.A. and A.L.P.O. (see p. 30).

Good handbooks for planet observers include, among others, Thomas A. Dobbins, Donald C. Parker, and Charles F. Capen, *Introduction to Observing and Photographing the Solar System* (Willmann–Bell, 1992), and Fred W. Price, *The Planet Observer's Handbook* (Cambridge, 1994). For current planetary research, see J. Kelly Beatty *et al.*, *The New Solar System* (Cambridge, fourth edition, 1999; revised regularly).

4.2 The view from Earth

The orbits of the planets are all close to the ecliptic, which is high in the evening sky in the spring and in the predawn sky in the autumn (in the Northern Hemisphere). Accordingly, those are the best times of year to observe planets.

The outer planets, those outside the Earth's orbit, are best seen at **opposition** (Figure 4.3), when the planet is closest to Earth and directly opposite the Sun; it then rises at sunset and culminates at midnight. During subsequent weeks

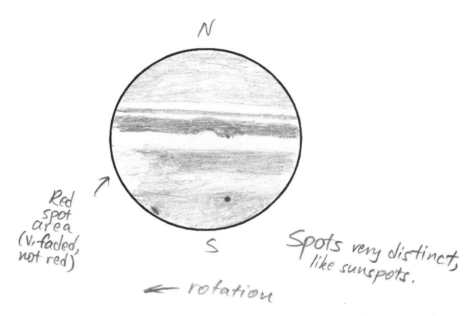

Figure 4.1. The author's pencil sketch of Jupiter shortly after pieces of Comet Shoemaker–Levy crashed into it. Drawn on 1994 July 21, 1:17–1:24 UT, with an 8-inch (20-cm) Schmidt–Cassegrain telescope at 222×. At the time, the South Equatorial Belt was somewhat faded and the Great Red Spot extremely so.

it moves into the evening sky. The phase of an outer planet is always full or gibbous; the gibbous phase is quite pronounced on Mars, but less on the more distant planets, whose orbits are much larger than the Earth's.

The inner planets, Mercury and Venus, go through a complete cycle of phases like the Moon (Figure 4.4). These phases, in fact, are what convinced Galileo that the planets must orbit the Sun rather than the Earth. If Mercury and Venus orbited the Earth, staying in line with the Sun but never passing completely around it, then they would be either always crescent or always gibbous.

When an inner planet is at the greatest apparent distance from the Sun, it is said to be at **greatest elongation**. At greatest eastern elongation, the planet is visible in the evening sky after sunset; at greatest western elongation, it is visible in the morning sky before sunrise. The apparent diameter of an inner planet changes appreciably with its position in its orbit. It is largest at inferior conjunction, when the planet cannot be seen.

Mercury and Venus are often observed in the daytime, when they are high in the sky. This is somewhat easier with computerized telescopes than with other types; you can level the tripod precisely, do a one-star alignment on the Sun, and then slew to Mercury or Venus and remove the solar filter. *Use extreme caution* to make sure that the telescope is not aimed at or near the Sun with the filter off. Mercury is visible, and Venus is quite easy to see, in a telescope at low power; first-magnitude stars can also be seen in the daytime at medium power and can be used to refine the alignment.

The planets

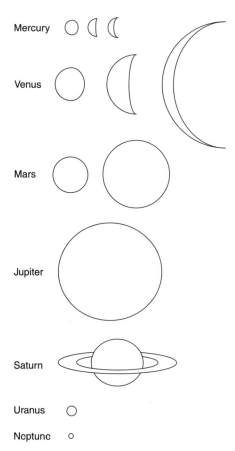

Mercury

Venus

Mars

Jupiter

Saturn

Uranus

Neptune

Figure 4.2. Relative sizes of the planets as seen from Earth. Some of them vary appreciably as they move in their orbits. (From *Astrophotography for the Amateur*.)

The time a planet takes to orbit the Sun, relative to the stars, is called its **sidereal period**. Its orbital period relative to the Earth is its **synodic period** and is the average interval at which oppositions or elongations recur.

4.3   Mercury

Diameter	4880 km (0.4 × Earth)
Rotation period	58.6462 days (2/3 × orbital period)
Apparent diameter (at typical elongation)	8″
Brightness (at typical elongation)	Mag. 0.5
Mean distance from Sun	57.6 million km (0.39 × Earth)
Orbital period	87.968 days
Elongations recur every	115.88 days
Maximum elongation from Sun	28°
Satellites	None

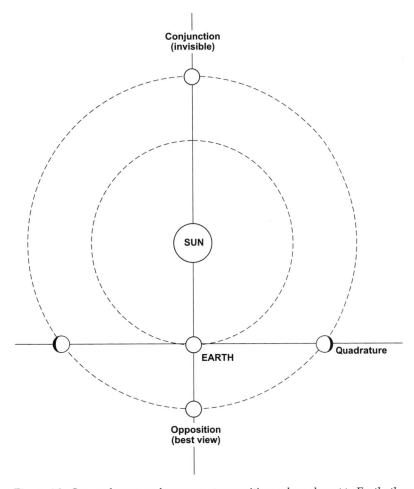

Figure 4.3. Outer planets are best seen at opposition, when closest to Earth; the phase is always full or gibbous.

4.3.1 Elongations of Mercury, 2002–2010

Because Mercury's orbit is so far from a perfect circle, its maximum elongation from the Sun ranges from 18° to 28°.

These and other planetary elongation and opposition dates are from tables published by Chris Marriott for use with *SkyMap Pro* (http://www.skymap.com).

Eastern (Evening) (best in spring)	Western (Morning) (best in autumn)
2002 Jan. 11 (19°)	2002 Feb. 21 (26°)
2002 May 04 (20°)	2002 Jun. 21 (22°)
2002 Sep. 01 (27°)	2002 Oct. 13 (18°)

2002 Dec. 26 (20°)	2003 Feb. 04 (25°)
2003 Apr. 16 (20°)	2003 Jun. 03 (24°)
2003 Aug. 14 (27°)	2003 Sep. 26 (17°)
2003 Dec. 09 (20°)	2004 Jan. 17 (24°)
2004 Mar. 29 (19°)	2004 May 14 (26°)
2004 Jul. 27 (27°)	2004 Sep. 09 (18°)
2004 Nov. 21 (22°)	2004 Dec. 29 (22°)
2005 Mar. 12 (18°)	2005 Apr. 26 (27°)
2005 Jul. 09 (26°)	2005 Aug. 23 (18°)
2005 Nov. 03 (24°)	2005 Dec. 12 (21°)
2006 Feb. 24 (18°)	2006 Apr. 08 (28°)
2006 Jun. 20 (25°)	2006 Aug. 07 (19°)
2006 Oct. 17 (25°)	2006 Nov. 25 (20°)
2007 Feb. 07 (18°)	2007 Mar. 22 (28°)
2007 Jun. 02 (23°)	2007 Jul. 20 (20°)
2007 Sep. 29 (26°)	2007 Nov. 08 (19°)
2008 Jan. 22 (19°)	2008 Mar. 03 (27°)
2008 May 14 (22°)	2008 Jul. 01 (22°)
2008 Sep. 11 (27°)	2008 Oct. 22 (18°)
2009 Jan. 04 (19°)	2009 Feb. 13 (26°)
2009 Apr. 26 (20°)	2009 Jun. 13 (23°)
2009 Aug. 24 (27°)	2009 Oct. 06 (18°)
2009 Dec. 18 (20°)	2010 Jan. 27 (25°)
2010 Apr. 08 (19°)	2010 May 26 (25°)
2010 Aug. 07 (27°)	2010 Sep. 19 (18°)
2010 Dec. 01 (21°)	

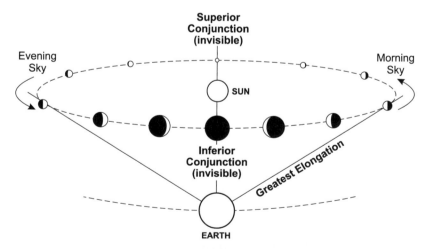

Figure 4.4. Looking across the Solar System at Mercury or Venus, we see the planet go through a full set of phases like those of the Moon.

4.3.2 Transits of Mercury

Mercury will pass across the visible face of the Sun on May 7, 2003 (about 5:15–10:30 UT, visible from Britain) and November 8, 2006 (about 19:15–0:15 UT, visible from the United States), and subsequently in 2016, 2039, 2049, and 2052.

Like occultations (p. 33), transits of Mercury provide an opportunity for accurate measurement of positions in the Solar System, and amateur timings are solicited by researchers.

4.3.3 Observing Mercury

Mercury is a dry, airless world, similar to our Moon and only slightly larger. The only space probe ever to visit Mercury was Mariner 10, which photographed about half of its heavily cratered surface in 1974. The craters are not visible from Earth, and although there are some streaks and patches like those of Mars, they are much fainter.

Because it is so close to the Sun, Mercury can only be observed in the daytime or low in the sky, and in either case the air is turbulent. Thus, observing Mercury is unusually challenging, and most amateurs will be able to do no more than follow the phases.

From the 1880s until 1962, telescopic observers thought, mistakenly, that Mercury always kept the same face toward the Sun. In fact, Mercury rotates one and a half times during each orbit. But if you observe Mercury only when it is north of the ecliptic – a practical necessity for a planet that is never high in the night sky – then you always see Mercury with the almost exactly the same side toward the Sun, at least for a few years. That is how E. M. Antoniadi and others managed to make consistent maps of Mercury despite the error.

4.4 Venus

Diameter	12 104 km (0.95 × Earth)
Rotation period	243.02 days
Apparent diameter (at typical elongation)	24″
Brightness (at typical elongation)	Mag. −4.3
Mean distance from Sun	108.21 million km (0.72 × Earth)
Orbital period	224.695 days
Elongations recur every	583.92 days
Maximum elongation from Sun	47°
Satellites	None

4.4.1 Elongations of Venus, 2002–2010

Each greatest elongation is 47° or slightly less.

Eastern (Evening)	Western (Morning)
2002 Aug. 22	2003 Jan. 11
2004 Mar. 29	2004 Aug. 17
2005 Nov. 03	2006 Mar. 25
2007 Jun. 09	2007 Oct. 28
2009 Jan. 14	2009 Jun. 05
2010 Aug. 20	

4.4.2 Transits of Venus

Venus will pass across the visible face of the Sun on June 8, 2004 (about 5:30–11:30 UT, visible from Britain) and June 5–6, 2012 (about 22:15–5:00 UT, visible from the western United States) after which there are no more transits of Venus until 2117.

4.4.3 Observing Venus

With an **albedo** (reflectivity) of 0.6, Venus is the whitest object in the Solar System. Because of its great brightness, Venus is a fine sight in the telescope, but its brightness gives rise to more than the usual number of optical illusions and mistakes. Reflections have been mistaken for satellites, and atmospheric turbulence on Earth masquerades as irregularities of Venus' terminator or limb.

Because Venus is surrounded by heavy clouds, no surface features are visible; we can only observe its upper atmosphere. "Dusky shadings" on Venus were long controversial, but they are now known to be quite definite in ultraviolet light, though rarely seen visually. Visual sightings are most likely when the observer is young (with eyes more sensitive to ultraviolet light) and the telescope is a reflector with a simple eyepiece (so that there is little UV-absorbing glass in the light path).

The observed phase of Venus does not agree exactly with the angle joining the Earth, Sun, and Venus; **dichotomy** (half phase) occurs when Venus should be slightly gibbous. This anomaly was discovered in 1793 by J. H. Schröter (Schroeter) [pronounced, approximately, *SHRUT-er*] and is called the **Schröter effect**. It is caused by the inherent fuzziness of the terminator falling on Venus' thick clouds. A weaker Schröter effect can sometimes be seen on Mercury or even the Moon, caused by the fact that the terminator area is weakly illuminated and blends into the sky background.

Also because of Venus' thick atmosphere, the horns of the crescent often extend to form more than half a circle. Near inferior conjunction, Venus has been photographed as a thin crescent extending into a complete, though thin, circle of light all the way around the planet.

More controversial is the **ashen light,** a faint luminosity of the non-sunlit portion of Venus, resembling earthshine on the Moon. Some observers see the ashen light regularly from Earth, but it has never been confirmed by space probes. It may be an illusion due to the tendency of the eye and brain to complete the circle, filling in the whole disk of Venus even when only part of the disk is seen. Or it may be due to light scattering, phosphorescence, or auroral activity on Venus.

More controversial yet is "negative visibility," in which the unlit part of Venus seems darker than the surrounding sky. This is almost certainly an illusion, but if real, it is due to silhouetting of Venus against the outer solar corona or interplanetary dust.

4.5   Mars

Diameter	6792 km (0.5 × Earth)
Rotation period	24.623 hours
Apparent diameter (at opposition)	18″ (varies widely)
Brightness (at opposition)	Mag. −2.0
Mean distance from Sun	227.94 million km (1.52 × Earth)
Orbital period	686.93 days
Oppositions recur every	779.94 days

4.5.1   Oppositions of Mars, 2002–2010

Date	Apparent diameter	Declination	Martian season at the time (in Mars' northern hemisphere)
2003 Aug. 28	25″	−15°	Late autumn
2005 Nov. 07	20″	+16°	Mid-winter
2007 Dec. 24	16″	+27°	Early spring
2010 Jan. 29	14″	+22°	Mid-spring

4.5.2   Surface features of Mars

Mars is rewarding for the telescopic observer. Its features have relatively high contrast, so even a beginner can see them; telescopes as small as 2 inches (5 cm) reveal some surface detail under the best conditions, and you get a sense of immediacy from knowing that what you see is the planet itself, not its atmosphere or surroundings.

The surface of Mars has a split personality. The craters, mountains, and canyons revealed by space probes are almost completely unrelated to the shadings visible from Earth. One vast canyon, the Valles Marineris complex, is faintly visible as a streak previously called Agathodaemon or Coprates, and Mars' largest volcano, Olympus Mons, is a prominent white spot previously known as Nix Olympica. Apart from that, the **albedo features** (shadings) consist of superficial dust and do not correspond to the topographic features.

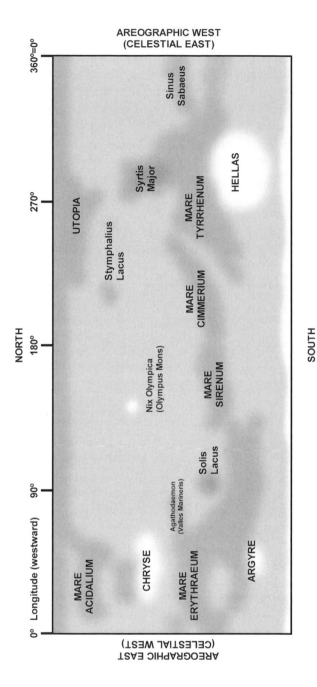

Figure 4.5. Mirror-image albedo map of Mars, matching the view through a telescope with a diagonal. (Most other maps are flipped top-to-bottom compared with this one.) The map covers only equatorial and temperate latitudes, to about ±60°.

Figure 4.5 can be compared directly to the view through the telescope, bearing in mind that the map does not cover the poles. The longitude of the central meridian (facing the Earth) can be obtained from observing handbooks or computer software. It is also useful to know the latitude of the center of the visible disk, since Mars' north or south pole is sometimes tilted appreciably toward us.

On the map, longitude is measured westward from 0° to 360°. This is the usual practice, but a few topographic maps measure longitude eastward from 0° to 360°, and some older maps go from 180° east to 180° west. All maps agree that the prime meridian (longitude 0°) is defined by a dark patch called Sinus Meridiani, between Sinus Sabaeus and Mare Erythraeum.

The so-called "canals" of Mars made famous by Percival Lowell a century ago are mere illusions. In 1877, Giovanni Schiaparelli [*skee-a-par-ELL-ee*] mapped a system of streaks that he called *canali*, which is Italian for "channels" but was translated into English as "canals." Lowell and his colleagues saw them as perfectly straight lines. That they are an illusion is obvious because on Lowell's drawings, the "canals" are straight even when near the limb, where straight lines on Mars would appear curved. Nonetheless, for a time the educated public believed the inhabitants of Mars had built canals to manage their scarce supply of water.

Though much thinner than ours, Mars' atmosphere is substantial enough to sustain dust storms, which conceal the albedo features and sometimes permanently alter them. Occasionally, a phenomenon called "blue clearing" occurs, during which the Martian atmosphere becomes unusually transparent and surface features are visible even through blue filters that would not normally show them.

Mars has white polar caps composed of a mixture of water ice and carbon dioxide. Each polar cap shrinks when it is turned toward the Sun (during Martian summer), and the surrounding dark features become darker and greener, a phenomenon once attributed to vegetation. Meanwhile, the opposite polar cap, turned away from the Sun, grows back to its original size. Ice crystals can apparently form in clouds anywhere on Mars; occasionally "flares" are seen when the Sun glints off them (*Sky & Telescope*, May, 2001, pp. 115–123).

Because the albedo features change somewhat from year to year, it can be interesting to make your own maps and globes of Mars. To make a globe, start with a cheap metal globe of Earth. Make pinpricks where the lines of latitude and longitude intersect so that you will be able to reconstruct the grid after repainting it white. Then paint it and plot the features in pencil or chalk.

4.5.3 Named Martian features

The following are the Latin names of the most prominent Martian features. These are "Continental" pronunciations (p. 5); alternatives are possible, and what is important is to accent the right syllable.

Mare	MAH-reh	
Acidalium	ah-sid-AHL-ee-um	Acidalian Sea (of the goddess Venus)
Australe	ows-TRAH-leh	Southern Sea
Boreum	BORE-ee-um	Northern Sea
Cimmerium	sim-MEHR-ee-um	Cimmerian Sea
Erythraeum	er-ith-RAY-um	Red Sea
Serpentis	sehr-PEN-tiss	Sea of the Serpent
Sirenum	sir-RAY-num	Sea of the Sirens
Tyrrhenum	tir-RAY-num	Tyrrhenian Sea
Lacus	LAH-kuss	
Niliacus	neel-EE-ahk-us	Lake of the Nile
Solis	SOLE-iss	Lake of the Sun
Stymphalius	stim-FALL-ee-us	Stymphalian Lake (in Greece)
Sinus	SEEN-us	
Aurorae	ow-ROAR-ay	Bay of the Dawn
Margaritifer	mar-ga-RIT-if-er	Pearl-Bearing Bay
Meridiani	mer-id-ee-AH-nee	Meridian Bay (longitude 0°)
Sabaeus	sa-BAY-us	Sabaean Bay (in Arabia)
Syrtis	SEER-tiss	
Major	MAH-yor	Larger Gulf (of Sidra, N. Africa)
Minor	MIN-or	Smaller Gulf (of Cabes, N. Africa)
Agathodaemon	ah-gah-tho-DAY-mon	Good Spirit
Argyre	AR-gear-ay	Silver Island
Chryse	KREES-ay	Golden Island
Coprates	COPE-rah-tace	Name of a Persian river
Hellas	HEL-las	Greece
Nix Olympica	NIX ol-IM-pick-a	Olympian Snow
Olympus Mons	o-LIMP-us MONS	Mount Olympus
Utopia	oo-TOE-pee-a	Never-Never Land (literally "no place")
Valles Marineris	VAHL-lace mah-rin-EHR-iss	Valleys of Mariner (the space probe)

Many of these names are ancient names of places on Earth; others are mythological. Newly discovered topographic features have names identifying the type of formation; thus Mare Acidalium is now Acidalia Planitia [*pla-NITT-ee-ah*] (Acidalian Plain), and Nix Olympica is Olympus Mons. For more about Martian nomenclature, including details of the mythological references, go to http://www.usgs.gov/mars.

4.5.4 Satellites of Mars

Name	Pronunciation	Orbital period	Magnitude
Phobos	FO-bos	7.6 hours	11.3
Deimos	DEE-mos	30.3 hours	12.4

Almost drowned in the glare of Mars itself, these satellites are just within reach of a good 8- to 12-inch (20- to 30-cm) telescope when the planet is hidden by an occulting bar at the focal plane of the eyepiece.

4.6   Jupiter

Diameter	142 984 km (11.2 × Earth)
Rotation period	10 hours (see text)
Apparent diameter (at opposition)	46″
Brightness (at opposition)	Mag. −2.7
Mean distance from Sun	778.3 million km (5.2 × Earth)
Orbital period	11.8565 years
Oppositions recur every	398.88 days

4.6.1   Oppositions of Jupiter, 2002–2010

Date	Declination
2002 Jan. 01	+23°
2003 Feb. 02	+18°
2004 Mar. 04	+7°
2005 Apr. 03	−4°
2006 May 04	−14°
2007 Jun. 05	−21°
2008 Jul. 09	−22°
2009 Aug. 14	−15°
2010 Sep. 21	−2°

4.6.2   Surface features of Jupiter

Observable more than half of every year, Jupiter is many amateur astronomers' favorite planet. Four of the satellites can be seen in binoculars; a 2-inch (5-cm) telescope begins to show the dark bands, and a 5-inch (12-cm) or larger telescope reveals a wealth of detail.

Jupiter is a thick ball of gases surrounding a heavy core, and the "surface" that we see is actually its upper atmosphere. Accordingly, the surface features are constantly changing, and the usual work of the observer is to chart them, both by drawing and by timing their transit across the central meridian.

Jupiter's equator is approximately parallel with the ecliptic, so we always see it more or less straight-on. The belts and zones (Figure 4.7) are usually adorned with ovals, bumps, "festoons" (diagonal streaks reaching from one belt toward another), and other irregularities. The South Equatorial Belt (SEB) varies in prominence; for a while in the mid-1990s it was almost invisible. The Great Red Spot also varies; it was bright red in the 1970s but in recent years it has been rather pale. Typically, the Great Red Spot strengthens when the SEB fades, and vice versa. The SEB is often double; the NEB, some times so.

The light-colored zones correspond to strong eastward wind currents, blowing in the same direction as the planet's rotation; the dark belts are winds in the opposite direction, rather like the prevailing winds in opposite directions at

Figure 4.6. Photocopy or trace this outline when drawing Jupiter. Polar diameter is 93.5% of equatorial diameter, a shape known in drafting as a 70° ellipse.

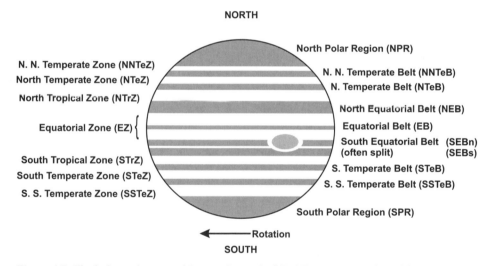

Figure 4.7. The belts and zones of Jupiter. Instead of 'Te,' Temperate is often abbreviated 'T'. Tropical is always 'Tr'. North is up and celestial west is to the left to match the view in a telescope with a diagonal.

different latitudes on Earth. The Great Red Spot is apparently a cyclonic storm that has been going on for more than 300 years.

The rotation of Jupiter is obvious to even the most casual observer (Figure 4.8), and drawings have to be finished quickly to prevent the features from being skewed. An alternative tactic is to make a "strip sketch" of a particular belt or zone, adding features to the drawing as they march past the central meridian.

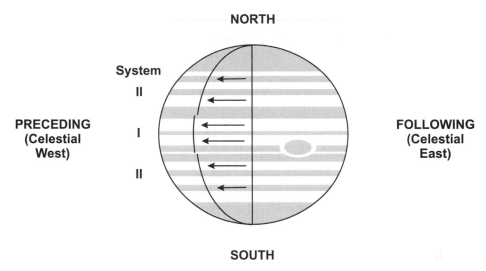

Figure 4.8. Jupiter rotates this far in one hour. System I (near equator) rotates slightly faster than System II. All planets rotate toward planetary east, celestial west, referred to as "preceding." The picture matches the view in a telescope with a diagonal.

The equatorial region (System I) rotates faster than the rest of the atmosphere (System II); the rotation periods are respectively 9 hours 50.5 minutes and 9 hours 55.7 minutes. There is also a System III, defined by radio observations of the core; its period is 9 hours 55.5 minutes, almost identical to that of System II. Each system has its own arbitrary scale of longitude.

4.6.3 Satellites of Jupiter

Name	Pronunciation	Orbital period	Magnitude	Apparent diameter
Io (I)	EYE-oh *or* EE-oh	1.77 days	5.0	1.2″
Europa (II)	yur-OH-pa	3.55 days	5.3	1.0″
Ganymede (III)	GAN-im-eed	7.16 days	4.6	1.8″
Callisto (IV)	ka-LIST-oh	16.69 days	5.6	1.6″

+ 12 named satellites fainter than magnitude 14

The four "Galilean" satellites (discovered by Galileo) put on a constant show as they orbit Jupiter. Their phenomena include **transits** across Jupiter's visible disk, **shadow transits** in which the shadow of a satellite falls on the planet, **eclipses** of the satellites in Jupiter's shadow, and, much less often, occultations or eclipses of one satellite by another. Predictions of these phenomena appear in the *Astronomical Almanac*, the *Handbook of the British Astronomical Association*, and *Sky & Telescope*.

Under good conditions, an 8-inch (20-cm) telescope shows each Galilean satellite as a tiny disk. Ganymede is the largest and has prominent light and dark patches that can be glimpsed with larger amateur telescopes. Callisto, second largest, is uniformly grayish, as is ice-covered Europa. Io, the closest to Jupiter of the four, is distinctly yellowish or orange.

The largest amateur telescopes may reveal two more satellites, Amalthea [*a-MALL-thee-a*], mag. 14, and Himalia [*him-ALE-ee-a*], mag. 14.8. Jupiter's thin, faint ring is beyond the reach of amateur equipment.

4.7 Saturn

Diameter	120 536 (9.4 × Earth)
Rotation period	10.6 hours (see text)
Apparent diameter (at opposition)	19″ (rings, 44″)
Brightness (at opposition)	Mag. 0.7
Mean distance from Sun	1429 million km (9.6 × Earth)
Orbital period	29.4235 years
Oppositions recur every	378.09 days

4.7.1 Oppositions of Saturn

Date	Declination	Inclination of rings
2002 Dec. 17	+22°	−27°
2003 Dec. 31	+22°	−26°
2005 Jan. 13	+21°	−23°
2006 Jan. 27	+19°	−19°
2007 Feb. 10	+15°	−14°
2008 Feb. 24	+11°	−8°
2009 Mar. 08	+6°	−3°
2010 Mar. 22	+2°	+3°

4.7.2 Surface features of Saturn

The surface of Saturn is similar to that of Jupiter but is much less active. Belts and zones, when visible, use the same nomenclature as for Jupiter. White spots are uncommon but should be observed carefully, since they provide an opportunity to measure the rotation period.

Like Jupiter, Saturn has different rotation periods at different latitudes. System I (10 hours 15 minutes) applies to the SEBn, EZ, and NEBs (see chart of Jupiter, Figure 4.7; the nomenclature is the same). System II (10 hours 38 minutes), now rarely used, was thought to be the rotation period of the rest of the globe. System III (10 hours 39.4 minutes) applies to radio emissions and is now also used for visual observations outside the equatorial region.

Saturn's equator is parallel to the ring plane, which in turn is not parallel to the ecliptic. Thus the angle from which we view the rings and the planet varies

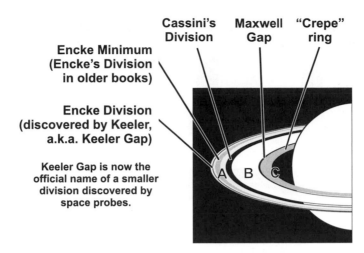

Figure 4.9. Nomenclature of Saturn's rings. The old term "Encke's Division" usually refers to the Encke Minimum (which is broad but indistinct) but sometimes refers to the Encke Gap (discovered by Keeler).

as Saturn goes around its orbit. Like Jupiter, Saturn is ellipsoidal in shape; its polar diameter is only 90% of its equatorial diameter.

4.7.3 Rings of Saturn

The rings of Saturn are a splendid sight in even the smallest telescopes. Even a 3-inch (7.5-cm) will show Rings A and B with Cassini's Division between them. Under good conditions, medium-sized amateur telescopes also show the Crepe Ring (Ring C), which is much fainter.

The rings are made of swarms of orbiting particles. Their structure is very complex and subject to change. In the past, observers reported a number of unusual stripes and spokes that were thought to be illusions but are now considered real. The ring system can even be asymmetrical, so if the rings seem brighter or thicker on one side of Saturn than the other, that fact should be recorded.

In 1837, J. F. Encke [pronounced *ENK-a*] discovered a broad dark stripe in the middle of Ring A, now known as the Encke Minimum and readily visible in amateur telescopes under good conditions.

Encke's name has also been attached to a very narrow but sharp division in Ring A discovered by James Keeler in 1888 with a 36-inch (90-cm) refractor; this is now known officially as the Encke Division, or occasionally the Keeler Gap. To further confuse us, the International Astronomical Union decided to honor Keeler by applying his name to an extremely thin division close to the outer edge of Ring A and visible only from space probes. Thus there are two Encke's Divisions and two Keeler Gaps, and the largest real division in Ring A goes by both names.

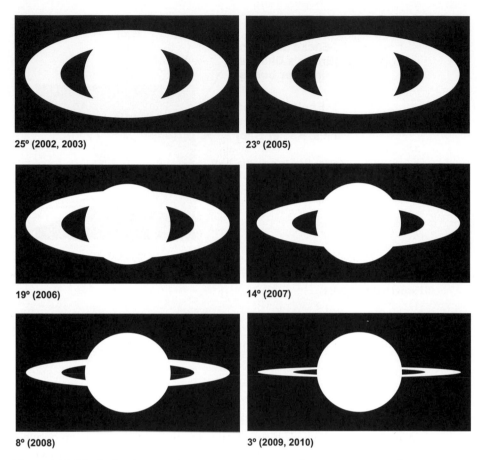

25° (2002, 2003)   **23° (2005)**

19° (2006)   **14° (2007)**

8° (2008)   **3° (2009, 2010)**

Figure 4.10. Blanks for drawing Saturn at various ring inclinations. Trace or photocopy, preferably somewhat enlarged.

Figure 4.10 can be photocopied to make blanks for drawing Saturn. On September 4, 2009, the Earth will be directly in line with the ring plane and the rings will be invisible.

4.7.4 Satellites of Saturn

Name	Pronunciation	Orbital period	Magnitude	Apparent diameter
Mimas	MIME-ahs	22.6 hours	12.9	
Enceladus	en-SELL-ad-us	1.37 days	11.7	
Tethys	TEETH-iss	1.89 days	10.2	
Dione	die-OH-nee	2.74 days	10.4	
Rhea	REE-a	4.52 days	9.7	
Titan	TIE-t'n	15.95 days	8.3	0.8″
Iapetus	eye-AP-et-us	79.33 days	10.2–11.9	

+11 named satellites fainter than magnitude 14

Instead of Jupiter's matched quartet, Saturn has a retinue of satellites of various sizes. Only Titan shows a disk in Earth-based telescopes, and even the largest telescopes show no surface detail. Iapetus has a large dark blotch on one side and is more than a magnitude fainter when east of the planet than when west of it.

The apparent orbits of the satellites are ellipses more-or-less parallel to the ring plane; the satellites' positions are given in standard handbooks and computer software.

4.8 Uranus

Diameter	51 118 km (4.0 × Earth)
Rotation period	17.24 hours
Apparent diameter (at opposition)	3.8″
Brightness (at opposition)	Mag. 5.5
Mean distance from Sun	2875 million km (19.2 × Earth)
Orbital period	83.7474 years
Oppositions recur every	369.66 days

4.8.1 Oppositions of Uranus

August 20, 2002; August 24, 2003; thereafter about 4 days later each year.

4.8.2 Surface features of Uranus

The rotation axis of Uranus is tilted 98°, so if "east" is the direction toward which Uranus rotates, then Uranus' "north" pole is 8° south of the ecliptic. For the next several years we will be seeing Uranus side-on, with its equator running roughly north–northwest to south–southeast in our sky.

Little detail is visible on Uranus in amateur telescopes, but a computerized mount makes the planet easy to find without a star map. Its blue color is quite distinctive. Under very good conditions Uranus can be seen with the naked eye.

4.8.3 Satellites of Uranus

Name	Pronunciation	Orbital period	Magnitude
Ariel	AIR-ee-ell	2.52 days	14.2
Umbriel	UM-bree-ell	4.14 days	14.8
Titania	tie-TANE-ee-a	8.71 days	13.7
Oberon	OH-ber-on	13.46 days	13.9

+13 named satellites fainter than magnitude 15

4.9 Neptune

Diameter	49 532 km (3.9 × Earth)
Rotation period	16.1 hours
Apparent diameter (at opposition)	2.5″
Brightness (at opposition)	Mag. 7.8
Mean distance from Sun	4 505 million km (30.1 × Earth)
Orbital period	163.72 years
Oppositions recur every	367.49 days

4.9.1 Oppositions of Neptune

August 2, 2002; August 4, 2003; thereafter about 2 days later each year.

4.9.2 Surface features of Neptune

Neptune is like Uranus, but more remote; surface features are unlikely to be seen in amateur telescopes, but observers with 10-inch (25-cm) and larger telescopes are urged to try, just in case a prominent spot develops. The equator of Neptune is inclined 30° to the ecliptic.

4.9.3 Satellites of Neptune

Name	Pronunciation	Orbital period	Magnitude
Triton	TRY-ton	5.88 days	13.5

+7 named satellites fainter than magnitude 15

4.10 Pluto

Diameter	2300 km (0.2 × Earth)
Rotation period	6.387 days
Apparent diameter (at opposition)	0.08″ (unobservable)
Brightness (at opposition)	Mag. 14
Mean distance from Sun	5915 million km (39.5 × Earth)
Orbital period	248.02 years (1.5 × Neptune)
Oppositions recur every	366.72 days

4.10.1 Oppositions of Pluto

June 7, 2002; June 9, 2003; thereafter 1 or 2 days later each year.

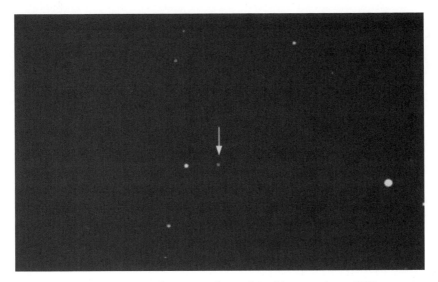

Figure 4.11. Pluto is just a 14th-magnitude speck in this two-minute CCD exposure taken with an 8-inch (20-cm) Schmidt–Cassegrain telescope at $f/6.3$. The bright star is SAO 160123, magnitude 10.

4.10.2 Telescopic appearance

For all practical purposes, Pluto seen from Earth is merely a starlike speck (Figure 4.11), visible in 8-inch (20-cm) and larger telescopes. The precise position of Pluto can be obtained from maps in the *Handbook of the British Astronomical Association* and *Sky & Telescope*. The computers built into telescopes do not always calculate orbits accurately enough to identify Pluto unambiguously, and although software packages such as *TheSky, Starry Night,* and *SkyMap Pro* give accurate coordinates, they do not show stars as faint as Pluto, so you have no way to distinguish the planet from the background stars.

I photographed Pluto by calculating the position with *TheSky,* then using high-precision mode to aim my Meade LX200 there, taking a CCD image, and comparing the results with an online Palomar Sky Survey image (see p. 99).

4.10.3 Satellite of Pluto

Name	Pronunciation	Orbital period	Magnitude
Charon	KAIR-on	6.39 days	16.8

Like the Earth and unlike any other planet, Pluto has a satellite not a great deal smaller than itself. Charon is just within reach of the largest amateur telescopes as a speck 0.9″ from Pluto and three magnitudes fainter.

Do not confuse Charon, which orbits Pluto, with the unusual comet/asteroid 2060 Chiron, which orbits the Sun slightly closer than Uranus.

Chapter 5
Comets, asteroids (minor planets), and artificial satellites

5.1 Small objects in the Solar System

A century ago, astronomers' picture of the Solar System was neat and clear. There were eight planets and an **asteroid belt,** a set of small, rocky bodies between the orbits of Mars and Jupiter, presumably the remains of a planet that had exploded or never formed.

By 1950 the picture was a bit less tidy. One planet, Pluto, was abnormally small and had an odd orbit. A few asteroids had been discovered with orbits outside the asteroid belt.

Today the situation is even less neat, and we are having to rethink old categories. Though the belt between Mars and Jupiter is predominant, there are asteroids all over the Solar System, including a second belt, the **Kuiper** (*KHOY-per*) **Belt**, outside the orbit of Neptune.

There is no longer a clear distinction between asteroids and comets; the asteroid 2060 Chiron has played both roles. It now appears that a comet is merely an icy asteroid from the outer Solar System that has been deflected close enough to the Sun to vaporize the ice.

The distinction between asteroids and planets is also becoming blurred; some scientists want to reclassify Pluto as the largest Kuiper-belt asteroid. Meanwhile, some planetary satellites, such as Mars' potato-shaped companions Phobos and Deimos, are physically indistinguishable from asteroids.

5.2 Orbits and ephemerides

What comets, asteroids, and artificial Earth satellites have in common is the fact that their orbits are subject to change. These small, lightweight objects are easily **perturbed** (deflected) by more massive bodies. Also, satellite orbits decay due to friction from the outermost layers of the Earth's atmosphere.

Accordingly, to observe them you must first compute their positions from accurate **orbital elements** (numbers that describe the orbit and the object's position in it). Fortunately, with personal computers and online data archives, this is now easy to do, and some telescopes (the Meade Autostar series) will even compute orbits from elements that you supply.

A table of the position of an orbiting body from day to day is called an **ephemeris** (*eff-EM-er-iss*), plural **ephemerides** (*eff-em-EHR-id-eez*). Ephemerides for the planets and a few asteroids are published in the *Astronomical Almanac* and the *Handbook of the British Astronomical Association*; the latter also covers a few periodic comets. Ephemerides for many more objects are available on the Internet. Orbits are explained in more detail beginning on p. 77.

5.3 Asteroids (minor planets)

5.3.1 Observing asteroids

An asteroid (or, perhaps more properly, a **minor planet**) is a small, rocky object that orbits the Sun. Most asteroid orbits are between the orbits of Mars and Jupiter and, like Mars, most asteroids are much closer to us at opposition than

Figure 5.1. Asteroids look like stars, hence their name, and can only be identified with an ephemeris or map. Here the asteroid 1 Ceres appears as an extra 7th-magnitude star in the Hyades cluster on November 17, 1998.

at other times. There are, however, asteroids all over the Solar System, and some of them occasionally pass close to Earth. One of them may eventually hit us, although no known asteroid is currently expected to do so.

The term **asteroid** means "starlike" and describes exactly how asteroids appear in a telescope; they are tiny points that never show any surface detail. The brightest asteroid, Vesta, reaches visual magnitude 6.5 and can be seen with the naked eye under ideal conditions. Hundreds of others reach at least magnitude 10 and are easy targets for amateur telescopes. The main way to distinguish an asteroid from a star is by its motion, which is typically 30″ per hour relative to the stars, but can be much more or much less depending on its position in its orbit.

Most amateur astronomers can do no more with asteroids than watch them. For the serious observer with a CCD camera, however, two kinds of scientifically useful observation are possible, **astrometry** (precise position measurement) and **photometry** (measurement of brightness, especially short-term variations). Precise positions are needed for many thousands of asteroids whose orbits are not yet known accurately. Brightness is of interest because many asteroids change brightness appreciably as they rotate, making it possible to infer the shape and rotation period of the asteroid.

Occultations of stars by asteroids are also of great interest, since the position, size, and shape of the asteroid can then be measured; observations are coordinated by the International Occultation Timing Association (p. 33).

5.3.2 Asteroid nomenclature and data

When the orbit of an asteroid has been determined accurately, the International Astronomical Union (IAU) gives the asteroid a number and invites the discoverer to propose a name. The sequence begins with 1 Ceres, discovered in 1801, and now extends past 26 000. Some memorable names include 2675 Tolkien, 13681 Monty Python, and 2309 Mr Spock – but frivolous names are now discouraged, and politically charged ones are prohibited.

Determining an accurate orbit can take ten years or more, so at a much earlier stage, as soon as a new asteroid's position has been measured on two nights, the IAU gives it a provisional designation. For instance, the first asteroid discovered in the first half of January 2009 will be called 2009 AA, the next one 2009 AB, and so on.

The first letter identifies the half-month in which the discovery took place. Twenty-four letters are used in sequence, A–H and J–Y; I and Z are not used. New letters start on the first and 16th of each month.

The second letter identifies the discovery itself. Twenty-five letters are used, the whole alphabet except I. After 25 discoveries during a single half-month, the second letter starts over at A_1; after 50 discoveries, A_2; and so on. In late September, 1998, there was such a good catch that the sequence reached 1998 SM_{165}.

Figure 5.2. Asteroid 1999 KW$_4$ swinging past Earth on May 26, 2001. It came within 0.032 AU (3 million miles, 5 million kilometers) and its apparent motion exceeded 1° per hour, but it was still no brighter than magnitude 10. Multiple exposure with a SBIG STV CCD camera, and an 8-inch (20-cm) Meade LX200 telescope, with a gap between the first two exposures.

The IAU Minor Planet Center maintains a web site at http://cfa-www. harvard.edu/iau/mpc.html with current data on comets and asteroids, including the ability to generate ephemerides on line. There is even a "Minor Planet Checker" that lets you input any right ascension or declination and check for asteroids in the vicinity. You can also input the NGC number of any galaxy and find asteroids that could be mistaken for supernovae in it.

5.3.3 Finding asteroids with computerized telescopes

One way to find an asteroid is to get its right ascension and declination from a published or online ephemeris, then go to those coordinates. Another way is to download the orbital elements into a telescope control program such as *TheSky, Starry Night,* or *SkyMap,* or even into the telescope itself, and let the computer do the finding.

The IAU web site provides orbital element files for many popular software packages and computerized telescopes, so do software manufacturers. I recommend going to the software manufacturer's web site first, so you can be sure that the file format has been verified to be correct for your software.

The Meade Autostar uses the file format shown in Figure 5.3; most others are similar though not identical. The files are always plain (ASCII) text and can be changed with a text editor or database program. After downloading a set of orbital elements, you should always calculate a few positions that you can verify independently, to make sure that none of the numbers were misidentified.

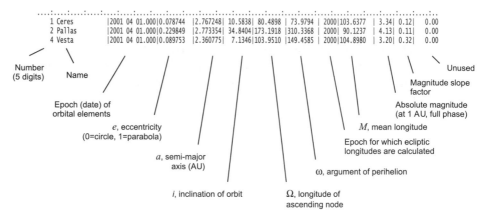

Figure 5.3. Meade Autostar file format for asteroid orbits; other software uses similar though not identical files. ASCII character | ("pipe") is used as a column separator. For explanation see p. 77.

Compared with comets or satellites, asteroid orbits do not change very fast, and orbital elements do not have to be extremely fresh to be usable. Elements that are two years old are generally fine. The details of the orbital elements are explained beginning on p. 77.

5.3.4 Discovering asteroids

The first asteroid was discovered in 1801; it took half a century to find five more. By 1991, there were 5000 numbered asteroids; today there are over 26 000.

The main thing that has sped up the pace of discovery is computer power. Anybody can *find* an asteroid; astronomers who search for novae and super-novae will tell you they've found too many! The hard part is *identifying* the asteroid and finding out whether it is already known. Today, the IAU's computers will do the job for you interactively through the Internet.

A second development is the use of CCD cameras by professionals since about 1980 and by amateurs since about 1990. Compared with film, CCDs are more sensitive and less affected by city lights. Also, they do not bend, stretch, or shrink, so positions measured from CCDs are always quite accurate.

For a while in the 1990s, amateurs with backyard telescopes could discover 17th-magnitude asteroids by taking CCD images and using the IAU's online ephemeris services. Today, Dennis Di Cicco, who pioneered amateur asteroid discovery, tells me that most asteroids down to magnitude 18 are picked up by the LINEAR and Spacewatch projects.

LINEAR is the Lincoln Near Earth Asteroid Research project, operated by the Massachusetts Institute of Technology with a 1-meter Cassegrain telescope in New Mexico. Its purpose is to detect objects that may collide with the Earth. Spacewatch is the University of Arizona asteroid research program. Each of them has found nearly half a million objects so far.

That doesn't mean amateur discoveries are impossible; neither LINEAR nor Spacewatch can catch everything in the whole sky, and asteroids with unusual orbits can be anywhere. Check the Minor Planet Center web site (see p. 65) for current advice on asteroid hunting. The ability to measure positions precisely relative to known stars is critical. There is also a great need for follow-up observations of known objects.

5.4 Comets

5.4.1 Observing comets

Most comets are discovered only a short time before they reach maximum visibility; thus, it is important to get news of discoveries quickly. The Internet is a big help here. I missed Comet West in 1975 because I was away from the astronomical news media at the time, and anyhow, the comet came and went before *Sky & Telescope* could get an account of it into print. Comet Hale–Bopp (C/1995 O1) was almost unique because, although very bright, it was discovered over two years before maximum brightness; thus everyone had plenty of time to get ready for it.

Figure 5.4. Comet Hale–Bopp (C/1995 O1) had separate ion and dust tails. A one-minute exposure on Kodak Elite Chrome 100 pushed to 200 speed, with a 100-mm lens at $f/2.8$, enhanced by slide duplication and digital processing. Comets as bright as this come along only a few times per century.

Today, the Internet keeps comet obsevers in touch. Like many amateur astronomers, I subscribe to weekly e-mail bulletins from *Sky & Telescope* and can check comet data online at numerous sites. Now my only excuse for missing a bright comet is cloudy weather.

A few comets have well-known periodic orbits – among them Halley's,[1] due back in 2062 – but most comets appear unexpectedly and seem to be making one-time visits to the inner Solar System. Astronomers believe most comets orbit the Sun peacefully in **Oort's cloud**, a region far beyond the orbit of Pluto that has not been observed directly. Occasionally a collision or close encounter deflects a comet toward the Sun, and as it warms up, its outer layers vaporize, forming a **coma** (gas cloud) and then, if the comet is a big one, a **tail** like those of Comets Hale–Bopp (C/1995 O1) or Hyakutake (C/1996 B2).[2]

Unlike those on TV, real comets do not streak across the sky, nor is the tail a trail that they leave behind. Comets move at about the same speed as planets, so a comet always appears to stand still in the sky, at least until you compare it very carefully with the background stars.

The tail is made of matter driven off by solar radiation and always points away from the Sun. Sometimes the white **dust tail** is distinguishable from the bluish **ion tail** of ionized gas molecules. Occasionally there is an **antitail**, a small jet pointing toward the Sun.

Comets are brightest when close to the Sun. That means they are often visible only in twilight, after sunset or before sunrise as the case may be. Thus, the serious comet observer will want a dark site with clear east and west horizons. Comets look a lot like nebulae or distant star clusters, and the observing techniques are the same – low power and good dark adaptation are needed. Most 4th- to 10th-magnitude comets look like faint galaxies or globular clusters.

5.4.2 Comet designations

Comets are named for their discoverers. Project LINEAR (p. 66) finds so many comets that there is usually at least one Comet LINEAR in the sky, sometimes two or three. Project Spacewatch also finds comets, but they are usually named after the members of the discovery team.

The official designations assigned by the International Astronomical Union (IAU) start with P/ for periodic comets, C/ for comets that appear to be making a one-time visit, D/ for defunct comets (such as the one that crashed into Jupiter in 1994), X/ for comets with unknown orbits, and A/ for comets that turn out to be asteroids.

If a periodic comet has been observed at more than one **perihelion** (close approach to the Sun), its prefix is preceded by a number and followed by

---

[1] *Halley* rhymes with *Sally*, not *Bailey*.

[2] In case you're still wondering, this one is pronounced *HYAH-koo-TAH-keh*. The *y* is a consonant.

Figure 5.5. A typical telescopic comet: C/2001 A2 (LINEAR) at 7th magnitude, with a short tail. The bright star is 1 Pegasi (magnitude 4). The central portion of a twelve-minute piggyback exposure on Elite Chrome 200 film with a 180-mm lens at $f/2.8$. The field of the picture is about 5° wide.

the discoverer's name: thus 1P/Halley, 2P/Encke, 109P/Swift–Tuttle. There are only 151 numbered comets.

Otherwise, the comet is officially known by a provisional designation like those for asteroids (p. 64) but with a number instead of the second letter. Thus the comets discovered in the second half of January 2009 will be C/2009 B1, C/2009 B2, and so on. These designations are normally followed by the name(s) of the discoverer(s) in parentheses, such as C/1995 O1 (Hale–Bopp).

Before 1995, the system was simpler. Comets were usually referred to by the discoverer's name, preceded by P/ if the comet was periodic (such as P/Halley). Two designations of a more technical type were also used. As soon as each comet was discovered, it received a designation such as 1982a, 1982b, etc., continuing after 1982z with $1982a_1$. As each comet passed perihelion it was given a Roman numeral designation. Thus Halley's Comet was designated 1982i when observers first recovered it and 1986 III when it reached perihelion four years later.

5.4.3 Finding comets with computerized telescopes

The *Handbook of the B.A.A.* gives ephemerides of a few periodic comets. Most comets, though, are unexpected visitors and you must download up-to-date elements from the Minor Planet Center (see p. 65) or the maker of your software.

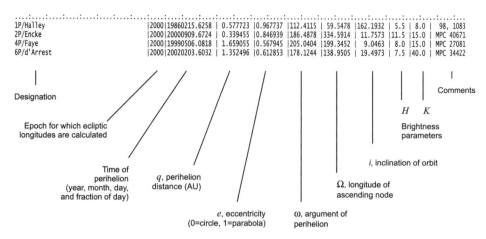

Figure 5.6. Meade Autostar file format for comet orbits; other software uses similar though not identical files. Compare with Figure 5.3. For an explanation see p. 77.

Note that *most comet orbits are not very accurate* because the comet has only been observed for a short time, so even with the best available orbital elements, you may have to search for a few degrees around the predicted position.

Figure 5.6 shows the file format used by the Meade Autostar; others are similar. Note that comet elements are rather different from minor planet elements because the entire orbit is not known, and *e* (eccentricity) is usually rather high; when *e* = 1 the orbit is indistinguishable from a parabola, which is another way of saying that although it may well be an ellipse, the length of the ellipse is very large and completely unknown. Thus, the size and orientation of the whole orbit cannot be given, and instead one works from the date and position of perihelion.

5.4.4 How to discover a comet

Discovering a comet is still the easiest way for an amateur astronomer to become famous. In 1965, Kaoru Ikeya, a Japanese amateur, got his name in every newspaper on Earth when his comet turned out to be a spectacular sight.

Even a 4-inch (10-cm) telescope is sufficient. All you have to do is sweep the sky in a systematic way, moving in azimuth and altitude as if plowing a field, making the sweeps overlap slightly. The most convenient way to do this will depend on your mount.

Computerized telescopes are a great help because you can read out the right ascension and declination at any time. In fact, to distinguish comets from galaxies and other fuzzy objects, all you have to do is keep your telescope connected to a computer running a sky map program, and look at the map.

Discovering a comet takes 200 to 1000 hours of sweeping the sky, but the distribution is very irregular; some people sweep for a very long time with no results and then get several discoveries in rapid succession. Newly visible comets are usually near the Sun, so the twilight sky bears careful examination, especially before dawn, when there are fewer other people looking at it. But comets can appear anywhere in the sky, and they can be revealed suddenly by solar or lunar eclipses.

At the time of discovery, a comet usually looks like a small, faint ball of fuzz. Thus, some practice observing 10th-magnitude elliptical galaxies and faint globular clusters is good preparation. Any telescope with a wide field, at low power, is well suited to the task. If anything, larger telescopes are probably not as good because their true fields are narrower. For comet searching, you need a dark country sky, clear horizons, and a great deal of patience.

5.4.5 Reporting a comet discovery

Once you've found your comet, what should you do? *Hesitate.* Pause and recheck everything. Caution is in order, because it's very likely you've found a galaxy, nebula, star cluster, or a comet that is already known. In 1969, I "discovered" Comet Tago-Sato-Kosaka, which was several degrees away from its published position. Fortunately, I caught the error before reporting it.

Note the exact time and the position of the object. The readout from your telescope or computer may be inaccurate, so sketch the comet's position among the stars in the field, identify the stars positively, and check the field against *at least* one major star atlas such as *Uranometria 2000.0*. Rule out reflections in your telescope, faint groups of stars, and even small clouds. If you can take photographs or CCD images, do so – repeatedly.

Within half an hour, you should see that the comet has moved slightly relative to the stars. If no motion is evident, the object is probably not a comet.

Check your discovery against the comet ephemerides at the Minor Planet Center (see p. 65). Meanwhile, try to get another observer to confirm your discovery – preferably an experienced amateur in your area, or else a nearby observatory. In Britain, discovery reports can be phoned to *The Astronomer* magazine at +44 (0) 1256 471074 (see also http://www.theastronomer.org/discoveries.html).

Finally, if you're *quite* sure you've found a new comet, e-mail a detailed report to the IAU at cbat@cfa.harvard.edu. The IAU strongly prefers that you get a confirming observation on a second night, or at least a very good confirmation by a second observer, before doing so. About 90% of the comet discoveries reported to them turn out not to be real.

> **A word to the wise:** Do *not* announce your discovery on the Internet, nor reveal any details to local news media, before it has been checked by experts. If someone misunderstands your observation and spreads the word in distorted form, you can quickly end up looking like a crackpot. Do not even use the Internet to appeal for confirmations; there's too much risk of getting untrustworthy helpers. Someone else may claim falsely to have discovered your object before you did.

For more good advice about discovering and observing comets, see *Observing Comets, Asteroids, Meteors, and the Zodiacal Light,* by Stephen J. Edberg and David H. Levy (Cambridge, 1994).

5.5 Meteors

Meteors are fragments of cometary or asteroidal material that get into the Earth's atmosphere and are burned up by air resistance. They are not normally observed with telescopes, of course, since they sweep rapidly across the sky. A particle the size of a grape produces a very memorable fireball; the typical visible meteor is more like a grain of sand. Telescopic meteors do exist; they are perfect miniatures of ordinary meteors, perhaps sixth magnitude, with trails half a degree long, and anyone who has used binoculars regularly has probably seen one.

Many meteors are **sporadic** (random), but many more are organized into showers traceable to clouds of debris that the Earth encounters at specific points in its orbit (Table 5.1). Thus each shower recurs at the same time each year, and

Table 5.1. *Major annual meteor showers*

Date (every year)	Name	Typical max. hourly rate	Remarks
Jan. 1–5	Quadrantids	40	Brief maximum on January 3 or 4 Radiant near θ Boötis
Apr. 21–23	Lyrids	15	Brief, variable shower
May 1–8	Eta Aquarids	20	Maximum around May 5
Jul. 15–Aug. 15	Delta Aquarids	20	Maximum July 27/28
Jul. 25–Aug. 18	Perseids	40–100	Reliable; occurs every year Maximum August 12
Oct. 16–26	Orionids	25	Maximum October 21
Nov. 15–19	Leonids	varies	Maximum November 18 Recurs in 33-year cycle Spectacular 1933, 1966, 1999–2001; due again around 2034
Dec. 11–17	Geminids	75	Maximum December 14 Usually a fine shower Radiant is high in evening sky

Source: From *Astrophotography for the Amateur,* updated.

the meteors seem to radiate from a point (the **radiant**) corresponding to the direction from which the particles approach us.

Meteor storms like the Leonids of 1966 and 2001 are rare; normally, a good shower produces a visible meteor every five or ten minutes. More meteors are seen after midnight, when the radiant is high in the sky and the sky above us is on the leading side of the Earth's motion around the Sun. An exception is the Geminid shower, which puts on a good show before midnight because of the direction from which the meteors are approaching.

5.6 Artificial Earth satellites

5.6.1 Observing satellites

On any clear evening, ten or twenty artificial satellites can be seen moving slowly across the sky. They look like 2nd- to 5th-magnitude stars and take perhaps five or ten minutes to move from horizon to horizon. They can move in roughly straight lines in any direction. Often they disappear in mid-sky as the satellite enters the Earth's shadow. A few, such as the Hubble Space Telescope and the International Space Station, are distinctly brighter than the rest. And when the Sun glints ("flares") off a shiny Iridium communications satellite, the effect is spectacular, like a slow, bright meteor (Figure 5.7).

Until recently, amateur astronomers took little interest in artificial satellites because there was no easy way to identify them or predict their positions. That has changed, thanks to computers, the Internet, and computerized telescopes. Today you can go to http://www.heavens-above.com and get detailed predictions of visible satellite passes, complete with star charts.

Meade Autostar telescopes can track satellites directly; other telescopes can do so when connected to a computer running *Satellite Tracker* (http://sattracker.hypermart.net), *C-Sat* (http://www.skyshow.com/csat), or other software packages.

5.6.2 Satellite orbits

Most, but not all, satellite orbits are nearly circular. Some orbits are parallel to the Earth's equator; others go over the poles; and still others are inclined at all angles in between.

The time that a satellite takes to orbit the Earth depends on its altitude. The lowest orbits, about 300 km up, have a period of about 90 minutes. The exact relation is:

$$\text{Period (minutes)} = 84.5 \times \left[\frac{6378 + \text{altitude (km)}}{6378} \right]^{3/2}$$

Figure 5.7. An "Iridium flare" – the Sun glinting off an Iridium communications satellite. Future plans for the Iridium satellite network are uncertain.

This is just Kepler's third law, which says that the period of an orbit is proportional to the 3/2 power of its radius; 6378 km is the radius of the Earth.

At an altitude of 35 900 km, a satellite orbits the Earth once every 24 hours. If such a satellite is directly above the equator and moving east, it is **geostationary** – that is, its orbital motion matches the rotation of the Earth, and it stays above the same point all the time. Many communications satellites are geostationary so that they can be "tracked" with antennas that don't move.

The brightness of a satellite is hard to estimate, but here is a very rough guide. If the average satellite 1000 km away is fourth magnitude, then for typical satellites at other distances:

$$\text{Magnitude} = 4 + 5\log_{10}\frac{\text{distance (km)}}{1000}$$

Thus, geostationary satellites are eleventh or twelfth magnitude, just within reach of amateur telescopes.

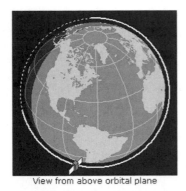
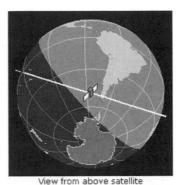

View from above orbital plane   View from above satellite

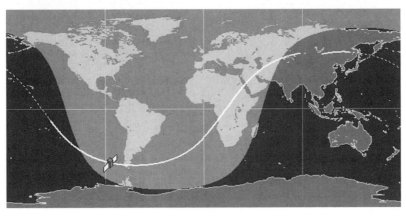

Ground track

Figure 5.8. Three views of an orbit of the International Space Station, inclined 51° to the Earth's equator. From www.heavens-above.com; reproduced by permission.

5.6.3 Satellite data files

Fortunately, there is just one standard file format for satellite orbits. It is the TLE (two-line-element) format from NORAD (the North American Aerospace Defense Command). Figure 5.9 describes it briefly. The most important parameter is the date (epoch), since that tells you how fresh the elements are, and for artificial satellites, *out-of-date elements give very inaccurate positions*. The date is given as a year and day number; 02333.5 is year 2002 on the 333rd day at 12^h UT. Fresh elements generally have an epoch slightly in the future of the day they are issued, to allow for gradual changes known to be taking place.

The second most useful parameter is the rate of motion, given as revolutions per day, since this tells you whether the satellite is low and therefore bright. The lowest satellites make about 16 revolutions per day.

Current TLE files can be downloaded from http://www.celestrak.com, http://oig1.gsfc.nasa.gov, and many other places; to find more, do a web search for "satellite tracking" or "two-line elements."

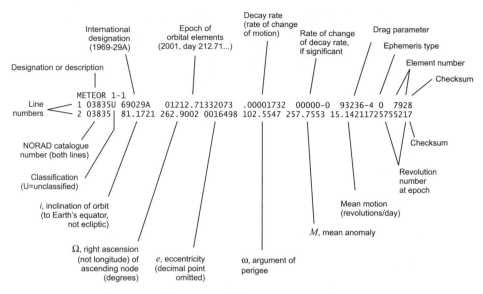

Figure 5.9. Standard TLE (two-line-element) format for artificial satellite orbits. For explanation see p. 77.

It goes without saying that whenever a spacecraft performs any orbital maneuver, all previous orbital elements become invalid. That's why the Space Shuttle is relatively hard to follow. NASA maintains a mailing list (accessible from http://spacelink.nasa.gov) to distribute updated elements quickly.

5.6.4 What to expect at the telescope

My personal feeling is that tracking satellites with a computerized telescope is more of a computational *tour de force* – and a stiff test of the motors – than a serious observing technique. After all, there is almost nothing to see. Only a few of the largest satellites, such as the International Space Station, can show anything more than a starlike point. But I am prepared to be proved wrong, and Figure 5.10 may be the first step toward doing so.

Satellite tracking certainly is not effortless. Downloading the most current orbital elements is only the first step. Regardless of what software you use, it is a good idea to start by going to http://www.heavens-above.com and getting accurate predictions.

My first attempt to view a satellite telescopically failed because the Meade Autostar, although happy to compute the orbit, does not keep track of whether the satellite is in the Earth's shadow; thus my ETX-90 tracked, or tried to track, an invisible object. On several subsequent tries the satellite either grazed the field of view or missed it altogether. It is a good idea to start with a relatively bright satellite that you can see with the unaided eye, so that if the telescope is mispointed, you can correct it.

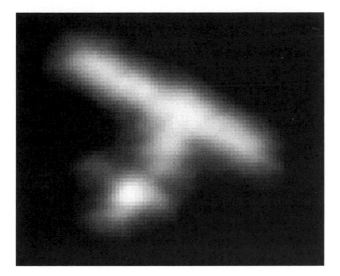

Figure 5.10. The Space Shuttle Atlantis docked to the International Space Station, imaged by Ulrich Beinert with a digital video camera and a Meade ETX-90 telescope on a non-computerized mount. Composite of two best video frames, digitally processed.

Remember that the observer's latitude, longitude, and local time must be entered into the software (or the telescope) *very* accurately (to within 0.1° and 1 second of time) and that only the freshest orbital elements can be used. To allow for a margin of error, the Autostar slews to a field that the satellite should pass through, then awaits your command to start tracking, and will pause again on command at any time.

5.7 Orbital elements explained

Orbits around the Sun or Earth are described in terms of **Keplerian elements** (numbers used in a calculation devised by Johannes Kepler, 1571–1630).

The orbit of one celestial object around another is always a circle, ellipse, parabola, or hyperbola (Figure 5.11). Of these, ellipses are by far the most common. Parabolas and hyperbolas are open orbits, in which one object swings by another and is deflected but does not revolve around it.

The elements specify the shape, size, and position of the orbit and the object's position on it. The *shape* requires just one parameter:

e the **eccentricity**, 0 for a perfect circle, 0.01 for a typical planetary orbit, 1 for a parabola, and greater than 1 for a hyperbola.

The *size* of the orbit is given either of two ways:

a the **semi-major axis** of the orbit, i.e., half the length of the ellipse, or else
q the **perihelion distance**, the minimum distance from the orbiting object to the central body. (For a parabola or hyperbola, $a = q$.)

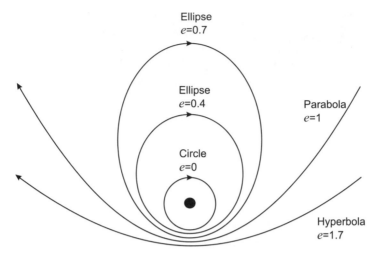

Figure 5.11. Possible orbits around a central mass. Eccentricity, e, describes shape.

Either of these is normally given in **astronomical units** (AU), where 1 AU is the mean distance from Earth to Sun (see p. 115).

The *position* of the orbit requires several more parameters:

i the **inclination** of the orbit relative to the ecliptic;

Ω the ecliptic **longitude of the ascending node,** where the plane of the orbit cuts through the plane of the ecliptic, moving northward; and

ϖ the ecliptic **longitude of the perihelion,** i.e., the point where the orbit comes closest to the central body, or

ω the **argument of the perihelion**, which is like ϖ but measured from Ω rather than from 0°.

Some old-fashioned symbols and terms deserve comment. Although it looks like ω with an accent, ϖ is actually a handwritten form of pi (π), an abbreviation for "perihelion" (a good Greek word). Capital omega (Ω) is a substitute for the medieval astrologers' symbol for ascending node (Ω), which is sometimes still used. "Argument" could at one time mean something like "increment."

All that remains is to specify where the object is in its orbit. For comets, this is done with one element:

T the time (and date) of perihelion.

For planets, position is usually given as something a bit more arcane:

L the **mean longitude,** the ecliptic longitude that the planet would have if it were moving in a uniform circle, or

M the **mean anomaly,** which is like L but measured from ϖ rather than from 0°.

These elements reflect Kepler's method of computing an ellipse by starting with a uniform circle and then correcting for e.

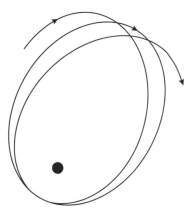

Figure 5.12. An orbit with rapidly changing ϖ due to the influence of a third body. Such things happen to asteroids and artificial satellites.

The term "anomaly" may puzzle you since there is certainly nothing unruly or anomalous about a longitude measured from ϖ. The term is a holdover from Ptolemy's ancient Earth-centered system. For Ptolemy, every orbit was the sum of two circles, a main circle and an epicycle. The longitude was the position on the main circle, and the "anomaly" was the position on the smaller circle, i.e., the deviation from the main circle, measured from ϖ of course. Copernicus moved the center of the system to the Sun, and Kepler got rid of the epicycles, but the name stuck.

Note that L and M are constantly changing as the planet moves, so the same orbit, described for different **epochs** (dates), will have completely different values of L and M.

For many elliptical orbits, ϖ and Ω are also constantly shifting (Figure 5.12). When they are not constant, their values at a particular time are called **osculating** ("temporarily touching") elements.[3]

Orbits of artificial Earth satellites are reckoned from the Earth's equator rather than the ecliptic, using right ascension (in degrees) rather than ecliptic longitude. Instead of the size of the orbit, the mean daily motion is given; either one can be calculated from the other.

To learn how to calculate positions from orbital elements, see *Practical Astronomy With Your Calculator*, by Peter Duffett-Smith (Cambridge, 1988). If you want to understand why orbits are elliptical, a good place to start is *Feynman's Lost Lecture*, by David and Judith Goodstein (Norton, 1996).

---

[3] Actually, "osculating" comes from the Latin for "kissing." Someone back there had a sense of humor!

Chapter 6
Constellations

6.1 Constellation names

The following 88 constellations are recognized by the International Astronomical Union and used on all modern star charts. The second form of each Latin name is the **genitive** (possessive), used after star designations, as in *Alpha Centauri* (Alpha of Centaurus).

These are "English Method" pronunciations (see p. 5). Other pronunciations are also widely used. See *Sky & Telescope,* May, 1997, pp. 68–69, and any large dictionary.

And	**Andromeda**	an-DROM-e-da	Princess Andromeda
	Andromedae	an-DROM-e-dee	(ancient)
Ant	**Antlia**	ANT-lee-a	The Air Pump
	Antliae	ANT-lee-ee	(Lacaille, 1763)
Aps	**Apus**	A-pus	The Bird of Paradise
	Apodis	A-po-diss	(Bayer, 1603)
Aqr	**Aquarius**	a-KWAY-ree-us	The Water-bearer
	Aquarii	a-KWAY-ree-eye	(ancient)
Aql	**Aquila**	ACK-will-a	The Eagle
	Aquilae	ACK-will-lee	(ancient)
Ara	**Ara**	AR-ah	The Altar
	Arae	AR-ee	(ancient)
Ari	**Aries**	A-ree-eez	The Ram
	Arietis	a-ree-EE-tiss	(ancient)
Aur	**Auriga**	aw-RYE-ga	The Charioteer
	Aurigae	aw-RYE-jee	(ancient)
Boo	**Boötes**	bo-OH-teez	The Herdsman
	Boötis	bo-OH-tiss	(ancient)

The dots (optional) indicate that the second *o* is a separate syllable.

| Cae | **Caelum** | SEE-lum | The Chisel |
| | **Caeli** | SEE-lye | (Lacaille, 1763) |

Coelum is an alternative Latin spelling no longer used.

Cam	**Camelopardalis**	kam-el-o-PAR-da-liss	The Giraffe
	Camelopardalis	kam-el-o-PAR-da-liss	(Bartsch, 1614)

Often called *Camelopardus* (gen. *Camelopardi*) prior to the IAU standard (1930).

Cnc	**Cancer**	CAN-ser	The Crab
	Cancri	CAN-cry	(ancient)
CVn	**Canes Venatici**	KAN-eez ven-AT-is-eye	The Hunting Dogs
	Canum Venaticorum	KAN-um ven-at-ik-O-rum	(Hevelius, 1687)
CMa	**Canis Major**	CAN-iss MAY-jor	The Larger Dog
	Canis Majoris	CAN-iss ma-JOR-iss	(ancient)

Some Latinists always write *j* as *i*: *Canis Maior(is)*.

CMi	**Canis Minor**	CAN-iss MY-nor	The Smaller Dog
	Canis Minoris	CAN-iss my-NOR-iss	(ancient)
Cap	**Capricornus**	kap-rick-CORN-us	The Goat
	Capricorni	kap-rick-CORN-eye	(ancient)
Car	**Carina**	ka-RYE-na	The Keel
	Carinae	ka-RYE-nee	(part of Argo Navis)
Cas	**Cassiopeia**	kass-ee-oh-PEE-ya	Queen Cassiopeia
	Cassiopeiae	kass-ee-oh-PEE-yee	(ancient)

Cassiopea and *Cassiepeia* are older Latin spellings no longer used.

Cen	**Centaurus**	sen-TAW-rus	The Centaur
	Centauri	sen-TAW-rye	(ancient)
Cep	**Cepheus**	SEE-fee-us	King Cepheus
	Cephei	SEE-fee-eye	(ancient)
Cet	**Cetus**	SEE-tus	The Whale
	Ceti	SEE-tie	(ancient)
Cha	**Chamaeleon**	ka-MEE-lee-on	The Chameleon
	Chamaeleontis	ka-mee-lee-ON-tiss	(Bayer, 1603)
Cir	**Circinus**	SUR-sin-us	The Drawing Compass
	Circini	SUR-sin-eye	(Lacaille, 1763)
Col	**Columba**	co-LUM-ba	[Noah's] Dove
	Columbae	co-LUM-bee	(Bayer, 1603)
Com	**Coma Berenices**	CO-ma ber-ee-NYE-seez	The Hair of Berenice
	Comae Berenices	CO-mee ber-ee-NYE-seez	(ancient)

Erroneously spelled *Coma(e) Berenicis* in Smyth's *Cycle of Celestial Objects*.

CrA	**Corona Australis**	ko-RO-na aws-TRAL-iss	The Southern Crown
	Coronae Australis	ko-RO-nee aws-TRAL-iss	(ancient)

Corona Austrina (gen. *Coronae Austrinae*) in some IAU publications.

CrB	**Corona Borealis**	ko-RO-na bo-ree-AL-iss	The Northern Crown
	Coronae Borealis	ko-RO-nee bo-ree-AL-iss	(ancient)
Crv	**Corvus**	KOR-vus	The Crow
	Corvi	KOR-veye	(ancient)
Crt	**Crater**	KRAY-ter	The Cup
	Crateris	kra-TEE-riss	(ancient)
Cru	**Crux**	kruks	The [Southern] Cross
	Crucis	KROO-siss	(originally part of Centaurus)

Cyg	**Cygnus** **Cygni**	SIG-nus SIG-nye	The Swan (ancient)
Del	**Delphinus** **Delphini**	del-FY-nus del-FY-nye	The Dolphin (ancient)
Dor	**Dorado** **Doradus**	doe-RAY-doe *or* dor-AH-doe doe-RAY-doos *or* dor-AH-doos	The Swordfish (Bayer, 1603)
Dra	**Draco** **Draconis**	DRAY-koe dra-KO-niss	The Dragon (ancient)
Equ	**Equuleus** **Equulei**	ek-WOO-lee-us ek-WOO-lee-eye	The Small Horse (ancient)
Eri	**Eridanus** **Eridani**	er-RID-an-us er-RID-an-eye	The River Eridanus (ancient)
For	**Fornax** **Fornacis**	FOR-naks for-NAY-siss	The Furnace (Lacaille, 1763)
Gem	**Gemini** **Geminorum**	JEM-in-eye jem-in-O-rum	The Twins (ancient)
Gru	**Grus** **Gruis**	groos GROO-iss	The Crane (Bayer, 1603)
Her	**Hercules** **Herculis**	HER-kyoo-leez HER-kyoo-liss	Hercules (ancient)
Hor	**Horologium** **Horologii**	hor-o-LOW-jee-um hor-o-LOW-jee-eye	The Clock (Lacaille, 1763)
Hya	**Hydra** **Hydrae**	HIGH-dra HIGH-dree	The Hydra (Monster) (ancient)
Hyi	**Hydrus** **Hydri**	HIGH-drus HIGH-dry	The Water Snake (Bayer, 1603)
Ind	**Indus** **Indi**	IN-dus IN-dye	The (American) Indian (Bayer, 1603)
Lac	**Lacerta** **Lacertae**	la-SER-ta la-SER-tee	The Lizard (Hevelius, 1687)
Leo	**Leo** **Leonis**	LEE-oh lee-OH-niss	The Lion (ancient)
LMi	**Leo Minor** **Leonis Minoris**	LEE-oh MY-nor lee-OH-nis my-NOR-iss	The Smaller Lion (Hevelius, 1687)
Lep	**Lepus** **Leporis**	LEP-us LEP-o-riss	The Hare (ancient)
Lib	**Libra** **Librae**	LYE-bra *or* LEE-bra LYE-bree *or* LEE-bree	The Scales (Balance) (ancient)
Lup	**Lupus** **Lupi**	LOO-pus LOO-pie	The Wolf (ancient)
Lyn	**Lynx** **Lyncis**	links LIN-siss	The Lynx (Hevelius, 1687)
Lyr	**Lyra** **Lyrae**	LYE-ra LYE-ree	The Lyre (ancient)
Men	**Mensa** **Mensae**	MEN-sa MEN-see	The Table [Mountain] (Lacaille, 1763)

Mic	**Microscopium** **Microscopii**	my-kro-SKO-pee-um my-kro-SKO-pee-eye	The Microscope (Lacaille, 1763)
Mon	**Monoceros** **Monocerotis**	mo-NOS-er-us mo-nos-er-OH-tiss	The Unicorn (Bartsch, 1614)
Mus	**Musca** **Muscae**	MUSS-ka MUSS-see *or* MUSS-kee	The Fly (Lacaille, 1763)
Nor	**Norma** **Normae**	NOR-ma NOR-mee	The Carpenter's [Level and] Square (Lacaille, 1763)
Oct	**Octans** **Octantis**	OCK-tans ock-TAN-tiss	The Octant (Lacaille, 1763)
Oph	**Ophiuchus** **Ophiuchi**	off-ee-OO-kus off-ee-OO-keye	The Snake-Holder (ancient)

Often called *Serpentarius* before 1800.

Ori	**Orion** **Orionis**	o-RYE-on o-rye-OH-niss	Orion the Hunter (ancient)
Pav	**Pavo** **Pavonis**	PAY-vo pa-VO-niss	The Peacock (Bayer, 1603)
Peg	**Pegasus** **Pegasi**	PEG-a-sus PEG-a-sigh	The Flying Horse (ancient)
Per	**Perseus** **Persei**	PER-see-us PER-see-eye	Perseus (ancient)
Phe	**Phoenix** **Phoenicis**	FEE-nix fee-NYE-siss	The Phoenix (Bayer, 1603)
Pic	**Pictor** **Pictoris**	PIK-tor pik-TOR-iss	The Painter['s Easel] (Lacaille, 1763)
Psc	**Pisces** **Piscium**	PIE-seez PIE-see-um	The Fishes (ancient)
PsA	**Piscis Austrinus** **Piscis Austrini**	PIE-siss aw-STRY-nus PIE-siss aw-STRY-nye	The Southern Fish (ancient)

Erroneously spelled *Pisces Austrinus* in Burnham's *Celestial Handbook*.
Piscis is singular; *pisces* is plural.

Pup	**Puppis** **Puppis**	PUP-iss *or* POOP-iss PUP-iss *or* POOP-iss	The Poop (The Stern) (part of Argo Navis)
Pyx	**Pyxis** **Pyxidis**	PICKS-iss PICKS-id-iss	The [Compass] Box (part of Argo Navis)
Ret	**Reticulum** **Reticuli**	ret-ICK-yoo-lum ret-ICK-yoo-lye	The Net or Reticle (Lacaille, 1763)
Sge	**Sagitta** **Sagittae**	sa-JIT-a sa-JIT-ee	The Arrow (ancient)
Sgr	**Sagittarius** **Sagittarii**	saj-it-AIR-ee-us saj-it-AIR-ee-eye	The Archer (ancient)
Scl	**Sculptor** **Sculptoris**	SKULP-tor skulp-TOR-iss	The Sculptor['s Studio] (Lacaille, 1763)
Sco	**Scorpius** **Scorpii**	SKOR-pee-us SKOR-pee-eye	The Scorpion (ancient)

Often called *Scorpio* (gen. *Scorpionis*) prior to the IAU standard (1930).

Sct	**Scutum**	SKYOO-tum	The Shield
	Scuti	SKYOO-tye	(Hevelius, 1687)
Ser	**Serpens**	SER-pens	The Snake
	Serpentis	ser-PEN-tiss	(ancient)

Consists of two separate regions on opposite sides of Ophiuchus:
Serpens Caput (gen. *Serpentis Capitis*) "the snake, the head" and
Serpens Cauda (gen. *Serpentis Caudae*) "the snake, the tail."

Sex	**Sextans**	SEX-tans	The Sextant
	Sextantis	sex-TAN-tiss	(Lacaille, 1763)
Tau	**Taurus**	TAW-rus	The Bull
	Tauri	TAW-rye	(ancient)
Tel	**Telescopium**	tel-es-KO-pee-um	The Telescope
	Telescopii	tel-es-KO-pee-eye	(Lacaille, 1763)
Tri	**Triangulum**	try-ANG-yoo-lum	The Triangle
	Trianguli	try-ANG-yoo-lye	(ancient)
TrA	**Triangulum Australe**	try-ANG-yoo-lum aw-STRAL-ee	The Southern Triangle
	Trianguli Australis	try-ANG-yoo-lye aw-STRAL-iss	(Bayer, 1603)
Tuc	**Tucana**	too-KAY-na *or* too-KAH-na	The Toucan
	Tucanae	too-KAY-nee *or* too-KAH-nee	(Bayer, 1603)

Called *Toucanus,* gen. *Toucani,* in some older books.

| UMa | **Ursa Major** | UR-sa MAY-jor | The Larger Bear |
| | **Ursae Majoris** | UR-see ma-JOR-iss | (ancient) |

Some Latinists always write *j* as *i: Ursa(e) Maior(is).*

UMi	**Ursa Minor**	UR-sa MY-nor	The Smaller Bear
	Ursae Minoris	UR-see my-NOR-iss	(ancient)
Vel	**Vela**	VEE-la	The Sails
	Velorum	vee-LO-rum	(part of Argo Navis)
Vir	**Virgo**	VUR-go	The Maiden
	Virginis	VUR-jin-iss	(ancient)
Vol	**Volans**	VO-lans	The Flying [Fish]
	Volantis	vo-LAN-tiss	(Bayer, 1603)
Vul	**Vulpecula**	vul-PEK-yoo-la	The Little Fox
	Vulpeculae	vul-PEK-yoo-lee	(Hevelius, 1687)

6.2   How the constellations got their names

The names of the constellations labeled "ancient" in the table are at least 2000 years old, and in some cases twice that old, or more. Because of precession, Centaurus, Crux, and Ara were visible from Europe in ancient times, but Sculptor and Fornax were not.

While plotting their pioneer star atlases, Johannes Bayer [pronounced *BY-er*] (1603) and Johannes Hevelius (1687) introduced new constellations in previously "unformed" areas. Bayer's constellations came from mariners' accounts of the southern sky. Hevelius invented his own; most of them are small animals, depicted in a happy flock joining him on the frontispiece of his atlas.

Most of the southern sky was unexplored until Nicolas Louis de Lacaille [*la-KYE*] observed it from South Africa in the mid-1700s and created many new constellations, mostly inanimate objects (even one honoring the eyepiece reticle with which he made his measurements).

In 1930 the International Astronomical Union (IAU) adopted an official list of 88 constellations with agreed-upon boundaries. The boundaries run along lines of right ascension and declination for epoch 1875; precession has skewed them relative to today's coordinates.

The constellation names are Latin; many are derived from Greek. *Tucana* is a Latinized form of a Native American word. *Dorado* (the Golden Fish or Swordfish) has the most dubious etymology; it is Spanish for "golden," from Latin *deauratus.* Not realizing its origin, Bayer took the word back into Latin and gave it a totally spurious genitive, *Doradūs.* So now we have a Latin word derived from a Spanish word derived from a Latin word – an etymological case of "I'm my own grandpa."

Not all Latin-named patches of sky are constellations. The **Hyades** [*HIGH-a-deez*] and **Pleiades** [*PLEE-a-deez*] are star clusters in Taurus. **Nubecula Major**

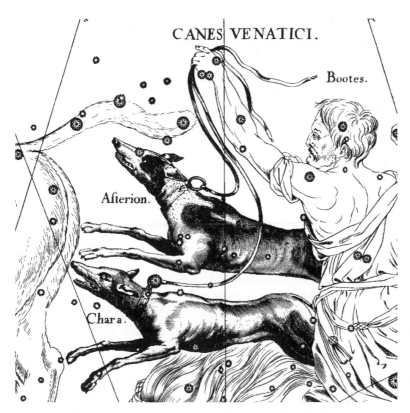

Figure 6.1. Hevelius invented the constellation Canes Venatici and named the dogs Asterion and Chara [*KAR-a*]. (*Firmamentum Sobiescianum*, 2nd edn, 1690.)

and **Nubecula Minor** are old names for the Magellanic Clouds, satellite galaxies of the Milky Way, in Dorado and Tucana respectively.

6.3 Obsolete constellations

In the official IAU list, the enormous southern constellation **Argo Navis** (the Ship Argo, gen. **Argūs Navis** with long *u*) has been broken up into Carina, Vela, Pyxis (formerly **Malus**, gen. **Mali**), and Puppis.

Also, many longer names have been shortened. For example, **Equuleus Pictoris** (the Painter's Easel) is now simply Pictor, and **Piscis Volans** (the Flying Fish) is just Volans. *Serpentarius* is an old alternative name for Ophiuchus.

Smyth's classic *A Cycle of Celestial Objects* (1844) treats the Pleiades as a constellation and uses four more constellation names that are not recognized today. **Anser** (the Goose) is now part of Vulpecula. **Antinoüs** (an-TIN-o-us), added in A.D. 130 by the Emperor Hadrian, is now part of Aquila. **Clypeus Sobieskii** (the Shield of Sobieski) is the constellation we now call Scutum. **Taurus Poniatowskii** is now part of Ophiuchus.

Eighteenth- and nineteenth-century maps introduced many other constellations that did not catch on. One of them, **Quadrans Muralis** (the Mural Quadrant, gen. **Quadrantis Muralis**), near Boötes, lent its name to the Quadrantid meteor shower that recurs every January. There was once another **Musca** (the Northern Fly) near Aries, and an owl, **Noctua**, near Libra. **Cervus** (the Deer) is another name for Monoceros.

6.4 The zodiac

Among the most ancient constellations are those of the **zodiac** ("belt of animals") through which the Sun, Moon, and planets pass during their orbital motions. These are, in order, **Aries, Taurus, Gemini, Cancer, Leo, Virgo, Libra, Scorpius, Sagittarius, Capricornus, Aquarius,** and **Pisces.**

Traditionally, the zodiac is divided into **signs** (segments) each exactly 30° long, starting with Aries at right ascension 0^h0^m. Because of precession, the signs no longer coincide with the constellations; the sign of Aries, for example, is now almost entirely within the constellation Pisces.

Until the 1600s, **astrology** (fortunetelling with horoscopes) was not entirely separate from astronomy; it was widely thought that the positions of the planets could influence human behavior in much the same way as the weather does.

Modern astronomers reject astrology for three reasons. First, no one has shown convincingly that any such influences actually exist. Second, no one has shown a mechanism by which such influences might work. (It isn't gravity or any known kind of radiation.) Third, and perhaps most importantly, astrologers today seem uninterested in overcoming these objections; they just practice their traditional craft, making predictions that are not experimentally testable.

Chapter 7
Stars – identification, nomenclature, and maps

7.1 Star names

7.1.1 Traditional names

The names in the following list are those recommended by the Yale *Bright Star Catalog* (YBS; see p. 101), plus all standard navigational stars, all the Greek names of the Pleiades, and a few other names that are used by computerized telescopes. You can safely assume that names not in this list are not commonly used.

The star positions are measurements by the Hipparcos satellite (epoch J2000.0). If a star is double, the position is that of the brighter member, but the magnitude is that of the two stars combined.

This list gives you *one* reasonably acceptable way to pronounce each name. There are many others. Many of the names are corrupted Arabic words that have not been pronounced correctly for centuries; any pronunciation that seems to fit the spelling is fair game.

You will probably never need to learn more than a few of these names, since there are other, better ways to identify stars (see p. 91).

Name			Mag.	R.A.	Dec. (2000.0)
Acamar	A-ka-mar	θ Eridani	2.9	$02^h58^m16^s$	$-40°18'17''$
Achernar	A-ker-nar	α Eridani	0.5	$01^h37^m43^s$	$-57°14'12''$
Acrux	A-kruks	α Crucis	0.8	$12^h26^m36^s$	$-63°05'57''$
Ad(h)ara	a-DARE-a	ϵ Canis Majoris	1.5	$06^h58^m38^s$	$-28°58'20''$
Albireo	al-BEER-ee-oh	β Cygni	2.9	$19^h30^m43^s$	$+27°57'35''$
Alcor	AL-core	80 Ursae Majoris	4.0	$13^h25^m14^s$	$+54°59'17''$
Alcyone	al-SIGH-oh-nee	η Tauri	2.9	$03^h47^m29^s$	$+24°06'18''$
Aldebaran	al-DEB-a-ran	α Tauri	0.9	$04^h35^m55^s$	$+16°30'33''$
Alderamin	all-der-RAM-in	α Cephei	2.5	$21^h18^m35^s$	$+62°35'08''$
Algol	AL-gol	β Persei	2.1–3.4	$03^h08^m10^s$	$+40°57'20''$
Alhena	al-HEN-a	γ Geminorum	1.9	$06^h37^m43^s$	$+16°23'57''$

Alioth	AL-ee-oth	ϵ Ursae Majoris	1.8	$12^h54^m02^s$	$+55°57'35''$
Alkaid	al-KADE	η Ursae Majoris	1.8	$13^h47^m32^s$	$+49°18'48''$
Alma(a)k (Almach)	al-MACK	γ Andromedae	2.1	$02^h03^m54^s$	$+42°19'47''$
Alnair (Al Na'ir)	al-NAIR	α Gruis	1.7	$22^h08^m14^s$	$-46°57'40''$
Alnath	al-NATH	= Elnath, below.			
Alnilam	al-NIGH-lam	ϵ Orionis	1.7	$05^h36^m13^s$	$-01°12'07''$
Alnitak	al-NIGH-tack	ζ Orionis	1.7	$05^h40^m46^s$	$-01°56'33''$
Alphard	al-FARD	α Hydrae	2.0	$09^h27^m35^s$	$-08°39'31''$
Alphekka	al-FECK-a	α Coronae Bor	2.2	$15^h34^m41^s$	$+26°42'53''$
Alpheratz	al-FEE-ratz	α Andromedae	2.1	$00^h08^m23^s$	$+29°05'26''$
Altair	al-TAIR *or* AWL-tair	α Aquilae	0.8	$19^h50^m47^s$	$+08°52'06''$
Ankaa	ANK-a	α Phoenicis	2.4	$00^h26^m17^s$	$-42°18'22''$
Antares	an-TAY-reez	α Scorpii	1.1	$16^h29^m24^s$	$-26°25'55''$
Arcturus	ark-TOUR-us	α Boötis	-0.1	$14^h15^m40^s$	$+19°10'57''$
Arneb	AR-nebb	α Leporis	2.6	$05^h32^m44^s$	$-17°49'20''$
Asterope	as-TEHR-o-pee	21 Tauri	5.7	$03^h45^m54^s$	$+24°33'16''$
Atlas	AT-lass	27 Tauri	3.6	$03^h49^m10^s$	$+24°03'12''$
Atria	A-tree-a	α Tri Aus	1.9	$16^h48^m40^s$	$-69°01'40''$
Avior	A-vee-or	ϵ Carinae	1.9	$08^h22^m31^s$	$-59°30'34''$
Bellatrix	bel-LAY-trix	γ Orionis	1.6	$05^h25^m08^s$	$+06°20'59''$
Betelgeuse	BAY-tell-jooz	α Orionis	0.4–1.3	$05^h55^m10^s$	$+07°24'25''$
	Also spelled *Betelgeux*.				
	Spelled *Beteigeuse* on German maps because of an early mistranscription.				
Canopus	ka-NO-pus	α Carinae	-0.6	$06^h23^m57^s$	$-52°41'44''$
Capella	ka-PEL-la	α Aurigae	0.1	$05^h16^m41^s$	$+45°59'53''$
Caph	KAFF	β Cassiopeiae	2.3	$00^h09^m11^s$	$+59°08'59''$
Castor	KASS-ter	α Geminorum	1.6	$07^h34^m36^s$	$+31°53'18''$
Celaeno	sell-EE-no	16 Tauri	5.5	$03^h44^m48^s$	$+24°17'22''$
Cor Caroli	KOR CAROL-eye	α C. Venaticorum	2.9	12^h56^m02	$+38°19'06''$
Deneb	DEN-ebb	α Cygni	1.3	$20^h41^m26^s$	$+45°16'49''$
Deneb Kaitos	DEN-ebb KAY-toss	= Diphda, below.			
Denebola	den-EBB-o-la	β Leonis	2.1	$11^h49^m04^s$	$+14°34'19''$
Diphda	DIFF-da	β Ceti	2.0	$00^h43^m35^s$	$-17°59'12''$
Dubhe	DUB-ee *or* DOO-bee	α Ursae Majoris	1.8	$11^h03^m44^s$	$+61°45'04''$
Electra	ee-LECK-tra	17 Tauri	3.7	$03^h44^m53^s$	$+24°06'48''$
Elnath	el-NATH	β Tauri	1.7	$05^h26^m18^s$	$+28°36'27''$
Eltanin	el-TAY-nin	γ Draconis	2.2	$17^h56^m36^s$	$+51°29'20''$
Etamin	et-AM-in	= Eltanin, above.			
Enif	EE-niff	ϵ Pegasi	2.4	$21^h44^m11^s$	$+09°52'30''$
Fomalhaut	FOE-mal-hawt	α Piscis Aust.	1.2	$22^h57^m39^s$	$-29°37'20''$
Gacrux	GAY-kruks	γ Crucis	1.6	$12^h31^m10^s$	$-57°06'48''$
Gienah	JEE-na	γ Corvi	2.6	$12^h15^m48^s$	$-17°32'31''$
Hadar	HAY-dar	β Centauri	0.6	$14^h03^m49^s$	$-60°22'23''$

Hamal	HAM-'l	α Arietis	2.0	$02^h07^m10^s$	$+23°27'45''$
Hyades	HIGH-a-deez	(star cluster in Taurus)			
Izar	EYE-zar	ϵ Boötis	2.4	$14^h44^m59^s$	$+27°04'27''$
Kaus Australis	KOSS aw-STRAL-iss	ϵ Sagitarii	1.8	$18^h24^m10^s$	$-34°23'05''$
Koc(h)ab	KO-kahb	β Ursae Majoris	2.1	$14^h50^m42^s$	$+74°09'20''$
Maia	MY-a	20 Tauri	3.9	$03^h45^m50^s$	$+24°22'04''$
Markab	MAR-kab	α Pegasi	2.5	$23^h04^m46^s$	$+15°12'19''$
Megrez	MEG-rez	δ Ursae Majoris	3.3	$12^h15^m26^s$	$+57°01'57''$
Menkar	MEN-kar	α Ceti	2.5	$03^h02^m16^s$	$+04°05'23''$
Menkent	MEN-kent	θ Centauri	2.1	$14^h06^m41^s$	$-36°22'12''$
Merak	MEE-rack	β Ursae Majoris	2.3	$11^h01^m50^s$	$+56°22'57''$
Merope	MEHR-o-pee	23 Tauri	4.1	$03^h46^m20^s$	$+23°56'54''$
Miaplacidus	my-a-PLASS-id-us	β Carinae	1.7	$09^h13^m12^s$	$-69°43'02''$
Mimosa	mim-O-sa	β Crucis	1.3	$12^h47^m43^s$	$-59°41'20''$
Mintaka	MIN-ta-ka or min-TACK-a	δ Orionis	2.3	$05^h32^m00^s$	$-00°17'57''$
Mira	MY-ra	o Ceti	2–10	$02^h19^m21^s$	$-02°58'40''$
Mirach	MERE-ack	β Andromedae	2.1	$01^h09^m44^s$	$+35°37'14''$

Same name has also been used for Izar, above.

Mirfak	MUR-fack	α Persei	1.8	$03^h24^m19^s$	$+49°51'40''$
Mizar	MY-zar	ζ Ursae Majoris	2.2	$13^h23^m56^s$	$+54°55'31''$

Same name has also been used for Izar (ϵ Boo) and Mirach (β And), above.

Navi	NAV-ee	γ Cassiopeiae	2.2	$00^h56^m43^s$	$+60°43'00''$
Nihal	NIGH-hal	β Leporis	2.8	$05^h28^m15^s$	$-20°45'34''$
Nunki	NUN-key	σ Sagitarii	2.0	$18^h55^m16^s$	$-26°17'48''$
Peacock	PEA-cock	α Pavonis	1.9	$20^h25^m39^s$	$-56°44'06''$
Phact	FACT	α Columbae	2.7	$05^h39^m39^s$	$-34°04'27''$
Phad	FAD	= Phact, above.			
Pleiades	PLEE-a-deez	(star cluster in Taurus)			
Pleione	plee-O-nee	28 Tauri	5.1	$03^h49^m11^s$	$+24°08'12''$
Polaris	po-LAIR-iss	α Ursae Minoris	2.0	$02^h31^m49^s$	$+89°15'51''$
Pollux	POL-lucks	β Geminorum	1.2	$07^h45^m19^s$	$+28°01'34''$
Procyon	PRO-see-ohn	α Canis Minoris	0.4	$07^h39^m18^s$	$+05°13'30''$
Pulcherrima	pull-KEHR-i-ma	= Izar, above.			
Rasalgethi	rahs-ahl-GETH-ee	α Herculis	2.8	$17^h14^m39^s$	$+14°23'25''$
Rasalhague	rahs-ahl-HA-gwee	α Ophiuchi	2.1	$17^h34^m56^s$	$+12°33'36''$
Regulus	REG-you-lus	α Leonis	1.4	$10^h08^m22^s$	$+11°58'02''$
Rigel	RYE-jel	β Orionis	0.2	$05^h14^m32^s$	$-08°12'06''$
Rigil Kent	RYE-gel KENT	α Centauri	−0.3	$14^h39^m36^s$	$-60°50'02''$
Sabik	SAY-bik	η Ophiuchi	2.4	$17^h10^m23^s$	$-15°43'30''$
Sadalmelik	sad-al-MELL-ick	α Aquarii	3.0	$22^h05^m47^s$	$-00°19'12''$
Saiph	SIFE or SAFE	κ Orionis	2.1	$05^h47^m45^s$	$-09°40'11''$
S(c)heat	SHE-at	β Pegasi	2.4	$23^h03^m46^s$	$+28°04'58''$

Schedar	SHED-ar	α Cassiopeiae	2.2	$00^h40^m30^s$	$+56°32'14''$
Shedir	SHED-er	= Schedar, above.			
Shaula	SHAW-la	λ Scorpii	1.6	$17^h33^m37^s$	$-37°06'14''$
Sirius	SEAR-ee-us	α Canis Majoris	-1.4	$06^h45^m09^s$	$-16°42'58''$
Spica	SPIKE-a	α Virginis	1.0	$13^h25^m12^s$	$-11°09'41''$
Sterope	STEHR-o-pee	= Asterope, above.			
Suhail	soo-HALE	λ Velorum	2.2	$09^h07^m60^s$	$-43°25'57''$
Tarazed	TAR-a-zed	γ Aquilae	2.7	$19^h46^m16^s$	$+10°36'48''$
Taygeta	ta-IJ-et-a	19 Tauri	4.3	$03^h45^m12^s$	$+24°28'02''$
Thuban	THOO-ban	α Draconis	3.7	$14^h04^m23^s$	$+64°22'33''$
Unukalhai	oo-NUCK-al-hye	α Serpentis	2.6	$15^h44^m16^s$	$+06°25'32''$
Vega	VEE-ga *or* VAY-ga	α Lyrae	0.0	$18^h36^m56^s$	$+38°47'01''$
	Spelled *Wega* in older books, from Arabic *Waki.*				
Vindemiatrix	vin-DEE-mee-A-tricks	ϵ Virginis	2.9	$13^h02^m11^s$	$+10°57'33''$
Zubenelgenubi	ZOO-ben-el-jen-OO-bee	α Librae	2.8	$14^h50^m53^s$	$-16°02'30''$

7.1.2 Other star names

About 3000 star names are given in the Yale *Bright Star Catalog* (YBS) and in *Star Names: Their Lore and Meaning,* by R. H. Allen (1899, reprinted by Dover, 1963). Most of these had fallen into complete disuse until the makers of computerized telescopes began to revive them.

The built-in catalogue of the Meade Autostar includes some names and comments from YBS that are puzzling out of context. For instance *Hyadum II* is δ^1 Tauri (SAO 93897), and *Miram in Becvar* [sic] is η Persei (SAO 23655). The latter was YBS's way of explaining that the star is called *Miram* in the *Atlas Catalogue* of Antonín Bečvář (1964).

The Meade LX200 refers to θ Aurigae as *Bogardus,* a name I have found nowhere else. Perhaps someone at Meade made it up to see if anyone else would use it.

In my opinion, the use of obscure star names should not be encouraged. The meanings of these names are often lost in the mists of history, the original spellings and pronunciations have been unknown for centuries, some names (e.g., *Mizar*) have been applied to more than one star, and some names of different stars are very similar (e.g., *Merak, Mirach*).

Instead, I recommend that amateurs follow the practice of professional astronomers and use standard star designations, such as Bayer letters (γ Cygni), Flamsteed numbers (37 Cygni) and catalogue designations (SAO 49528, HIP 100453).

7.1.3 Stars named after people

A few stars, such as Barnard's Star, are known by the names of astronomers who made important discoveries about them. These names are unofficial and are sanctioned only by common use.

In the late 1990s a number of companies advertised that, for a price, customers could name a star to honor a family member or personal hero. Many customers failed to realize that astronomers would never use these names or even hear about them. All the customer got was an elegantly printed certificate declaring that a star had been renamed – and in an alarming number of cases, the certificate did not adequately identify a star!

In at least two cases, people *have* succeeded in naming stars after themselves, at least temporarily. The eighteenth-century observer Nicolaus Venator smuggled his name into a star catalogue as Rotanev and Svalocin (α and β Delphini respectively; read them backward).

Continuing the backward-spelling game, astronauts Virgil Ivan Grissom, Roger Chaffee, and Edward H. White II managed to insert *Navi* (γ Cassiopeiae), *Regor* (γ Velorum), and *Dnoces* (ι Ursae Majoris) into a list of navigational stars for the Apollo space program. These names even appeared on charts in *Sky & Telescope*, and *Navi* is in the built-in catalogue of the Celestron NexStar. Sadly, Grissom, Chaffee, and White died in a fire in an Apollo command module being tested on the launch pad in January, 1967.

7.2 Modern star designations

7.2.1 Bayer and Lacaille letters

The "Alpha" in "Alpha Centauri" (α Centauri) is the work of Johannes Bayer (see p. 85), whose 1603 star atlas identified stars with lowercase Greek letters (Table 7.1). In each constellation, Bayer usually labeled the brightest star α, the second-brightest β, and so forth. After the 24 Greek letters, he used Roman letters a, b, c ... z, then A, B, C, and so on.

Computerized catalogues often use three-letter abbreviations for Greek letters alongside three-letter abbreviations for constellations (pp. 80–84). Thus α Centauri comes out ALP CEN, and θ^1 Eridani is THE1 ERI.

Table 7.1. *The Greek alphabet*

α	alpha	ι	iota	ρ (ϱ)	rho
β	beta	κ	kappa	σ	sigma
γ	gamma	λ	lambda	τ	tau
δ	delta	μ	mu	υ	upsilon
ϵ (ε)	epsilon	ν	nu	ϕ (φ)	phi
ζ	zeta	ξ	xi	χ	chi
η	eta	o	omicron	ψ	psi
θ (ϑ)	theta	π	pi	ω	omega

The alternate forms ε, ϑ, ϱ, and φ are more common on German charts.

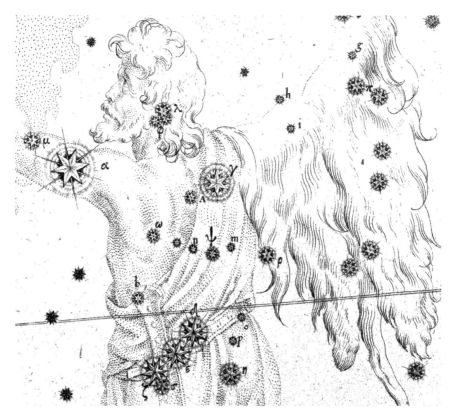

Figure 7.1. Northern Orion from Bayer's *Uranometria* (1603), with Greek and Roman letters identifying stars.

Bayer did not always follow a strict brightness sequence. Sometimes he followed chains of stars, and sometimes he labeled equally bright stars in an arbitrary order.

The Bayer letters were extended to the southern constellations by Lacaille (p. 85), who went all the way to Q in Centaurus, Puppis, and Vela. Note that many southern constellations have both an o (omicron) and an o, two different stars. Bayer letters are never italicized.

Letters from R to Z and letter pairs from AA to ZZ are variable-star designations (p. 137).

Other mapmakers have extended the Bayer/Lacaille designations in various ways; in fact no two star atlases match entirely! The mess is straightened out, as far as possible, in the Yale *Bright Star Catalogue* (see p. 101), which is considered authoritative.

Superscript numerals are particularly contentious. Sometimes they denote chains of stars, such as π^1–π^6 Orionis and τ^1–τ^9 Eridani. More often, they denote components of a double or multiple star, and it is unpredictable whether a particular catalogue will use them. Thus α^1 Centauri is another name for the star that double-star observers know as α Centauri A.

A few Bayer letters are in the "wrong" constellation because boundaries have changed or because a single star was part of more than one traditional constellation figure (e.g., α Andromedae = δ Pegasi). Even proper motion can disrupt star designations. In 1992, ρ Aquilae (SAO 105878) crawled across the border into Delphinus, making the designation ρ Aquilae invalid.

7.2.2 Flamsteed numbers

Designations such as "61 Cygni" come from the work of John Flamsteed (1725), who listed the stars of each constellation in order of increasing right ascension. (One of the stars he observed, 34 Tauri, turned out to be a prediscovery sighting of the planet Uranus.)

Today, Flamsteed numbers have supplanted all but the first few Bayer letters in each constellation.

The full set of Bayer/Lacaille letters is still used in the far southern sky, which Flamsteed could not see from England. In 1801, J. E. Bode (pronounced *BO-da*) published a star catalogue extending Flamsteed-style numbers to the south, but few astronomers used them; the only Bode numbers that survive are 47 Tucanae (actually a globular cluster) and 30 Doradus (actually a nebula).

7.2.3 STAR numbers

In various contexts you will encounter numbers such as "STAR 130." Unfortunately, these do not have any agreed-upon meaning. Usually they refer to the built-in catalogue of a particular telescope, such as LX200 or NexStar. STAR numbers also designate asterisms in Harrington, *The Deep Sky* (see p. 144).

7.3 Star maps

7.3.1 Wide-field atlases

When you're using a computerized telescope, you do not always need a star atlas to find objects. However, a wide-field atlas is still useful for learning the sky and planning observing sessions.

I recommend *The Cambridge Star Atlas*, by Wil Tirion (Cambridge University Press, 3rd edn., 2001). Its colorful charts are essentially an abridged version of *Sky Atlas 2000.0* (see below), with stars to magnitude 6.5 and a wide variety of deep-sky objects. Each chart is accompanied by a list of interesting objects. The atlas also contains a moon map and month-by-month charts of the whole sky.

The classic wide-field atlas is *Norton's Star Atlas*, originally published in 1910; the current edition is the 19th, edited by Ian Ridpath (Longman, 1998). The maps are in an older style with less emphasis on deep-sky objects but more labeling of faint stars and double stars. The accompanying reference material is very thorough.

7.3.2 Medium-scale atlases

A medium-scale atlas is one that goes somewhat beyond the naked-eye limit, so that it matches the view through a good finder. Atlases of this kind are especially useful for observing with binoculars and small wide-field telescopes (e.g., the ETX-60, ETX-70, NexStar 80) and for interpreting wide-field photographs.

The definitive atlas of this type is *Sky Atlas 2000.0,* by Wil Tirion and Roger W. Sinnott (Cambridge University Press, 1981, 2nd edn., 1998), which shows stars to 8th magnitude. Nebulae, clusters, and the Milky Way are plotted in detail. There are larger-scale charts of the Pleiades, the Virgo Galaxy Cluster, and other areas of interest.

This atlas follows the style of the classic *Atlas Coeli 1950.0 (Atlas of the Heavens),* by Antonín Bečvář of Skalnate Pleso Observatory, Czechoslovakia.[1] Published in 1948, Bečvář's was the first atlas to show the shapes of large nebulae accurately. Its color edition (1958) introduced the green nebulae, yellow clusters, and red galaxies with which observers are now so familiar.

7.3.3 Telescopic atlases

A telescopic atlas is one that you can compare directly to the view through the telescope.

In my opinion, every serious observer needs either a telescopic atlas or a software package to generate telescopic charts. A computerized telescope gets you to the neighborhood of an interesting object – but how do you identify the object itself? That's where charts are often essential.

For a long time, the only telescopic star atlas was the *Bonner Durchmusterung* of 1862 (see p. 101), and when astronomers needed detailed finder charts, they usually used photographs. Today, at least two good telescopic atlases are available.

Uranometria 2000.0, by Wil Tirion, Barry Rappaport, and Perry Remaklus (Willmann-Bell, 2nd edn., 2001) comprises two volumes that show stars to magnitude 9.75 and accurate outlines of nebulae in the style of *Sky Atlas 2000.0,* but without color or shading. The scale is $1° = 1.85$ cm. The second edition uses Hipparcos data and includes close-up charts of interesting regions to 11th magnitude, plus a smaller-scale atlas of the whole sky to magnitude 6.5 in 22 charts.

Millennium Star Atlas, by Roger W. Sinnott and Michael A. C. Perryman (European Space Agency, 1997) plots one million stars to magnitude 11 in three large volumes. It is the definitive plot of star positions measured by the Hipparcos and Tycho satellites (p. 101). It also shows thousands of deep-sky objects with accurate size and orientation, plus physical data about variable and binary stars. The scale is $1° = 3.6$ cm, so the typical low-power telescope field is more than an inch in diameter.

---

[1] The names are pronounced *AHN-toe-neen BETCH-varzh* and *SKAHL-nah-tay PLESS-o* respectively.

7.3.4 How to use a telescopic atlas

Using a telescopic atlas may be puzzling at first because, relative to the atlas, the image in the telescope is flipped left to right if the telescope has a diagonal, or upside down if it doesn't. Additionally, depending on the position of the telescope, the image may be tilted an unpredictable amount. The tilt is least when the telescope is aimed due south and the eyepiece is pointing straight up.

Figures 7.2 and 7.3 take you through the process of finding Barnard's Star, the nearest star other than the Sun that is visible from America and Europe (see p. 201).

Even with a computerized telescope, you can't go to Barnard's Star directly because it isn't in the built-in catalogue of most computerized telescopes. Instead, you'll find it by "star-hopping" from 66 Ophiuchi (SAO 123005, NexStar Star 4186), a 5th-magnitude star that you can find by computer or with the aid of a wide-field atlas. Meade LX200 users can shorten the journey by going directly to V566 Ophiuchi (GCVS 590566), a low-amplitude variable (mag. 7.5–7.9) with a period of just 10 hours.

Look first at Figure 7.2, a telescopic chart similar to *Uranometria* or the *Millennium Star Atlas*. As on all atlases, north is up and east is to the left.

The dotted lines highlight some important patterns in the stars. Atlases don't have these lines, of course; you will often draw such lines in pencil as you plan a star-hop.

Note that almost due south of 66 Ophiuchi is 67 Ophiuchi, a prominent double star. If you get lost, you can go to 67 Ophiuchi to confirm that you're in the right star field. Both 66 and 67 Ophiuchi are between 4th and 6th magnitude, easily visible in the smallest finder.

Right next to 66 Ophiuchi are two 9th-magnitude stars, joined to it by a bent dotted line. These three stars look like a miniature version of the constellation Aries. They will play a crucial role in this star-hop. In general:

- Angles, triangles, and bent rows of stars are easily recognizable when flipped and tilted. Individual stars, pairs, and straight lines are not.
- Three stars, not in a straight line, give you a basis for orientation. Two stars do not.

So this miniature Aries will be the beginning of the journey. At a 60° angle away from the outer star, I've drawn an arrow pointing toward Barnard's Star.

Note also that there is a prominent V-shaped group containing V566. If you can go from 66 Ophiuchi to this V-shaped group, then back up a third of the way, you'll be at your destination.

Two paths are possible. From the miniature Aries that surrounds 66 Ophiuchi, you can judge the appropriate angle (indicated by an arrow) and go straight to V566. Or you can go northwest to SAO 122981, then west to V566. The latter is more practical if you have an equatorially mounted telescope, which makes it easy to go straight north, east, west, or south. With an altazimuth mount, however, there is nothing special about those directions.

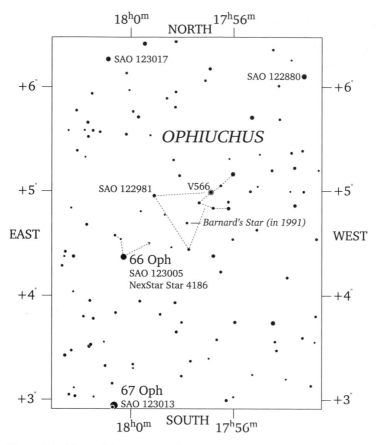

Figure 7.2. Chart of the position of Barnard's Star, showing stars to magnitude 9.8. Like an atlas, this chart has north at the top, east at the left. Dotted lines show a strategy for finding the star. (Hipparcos data plotted with *TheSky* software, copyright 2001 Software Bisque, Inc., used by permission.)

> **Hint:** When the telescope is on an altazimuth mount, the North, South, East, and West buttons on the keypad do *not* normally move in those directions.

Now look at Figure 7.3, which shows what you might actually see in a telescope with a diagonal at about 60×. The true field is slightly less than a degree in diameter, and circle 1 shows the field centered on 66 Ophiuchi. Proceed as follows:

- Find the two 9th-magnitude stars that form a miniature Aries.
- Make a 60° angle away from the outer star, just like the arrow on the map (even though it's flipped and tilted here).
- Move in that direction a distance equal to twice the field diameter, so that you're looking at the V-shaped group around V566 (circle 2).
- Now move back a third of the way to find Barnard's Star, which is in the middle of an equilateral triangle with V566 at one corner (circle 3).

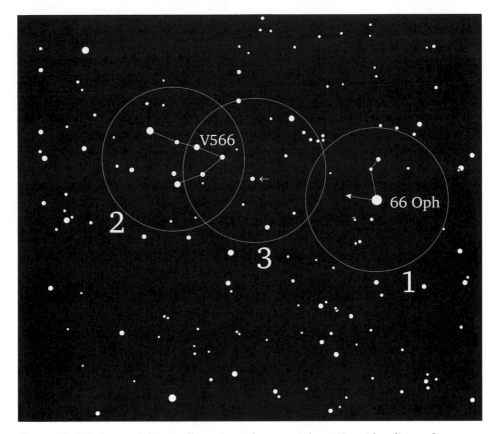

Figure 7.3. What you might actually see in a telescope at about 60× with a diagonal. Compared to the map, the image is mirror-imaged and somewhat tilted. The two stars next to 66 Ophiuchi tell you what direction to move to find the V-shaped group around V566; then back up and you're looking at Barnard's Star (arrow). (Adapted from Figure 7.2.)

Because of its high proper motion, Barnard's Star is already slightly north-northwest of the position measured by Hipparcos in 1991.

Star-hopping takes patience and practice, but with a computerized telescope, you can bail out and start over at any time. Just press GO TO again and you're back at 66 Ophiuchi.

7.3.5 Sky mapping software

Why rely on a printed star map when you can create one on your computer at any time? Computer programs such as *Starry Night* (http://www.starrynight.com), *TheSky* (Software Bisque, http://www.bisque.com), and *SkyMap* (http://www.skymap.com) can plot the stars with any scale and magnitude limit. You can customize maps to show the field of view of your eyepiece and to flip the image if you are using a diagonal.

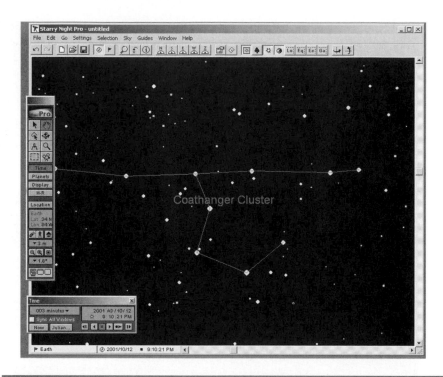

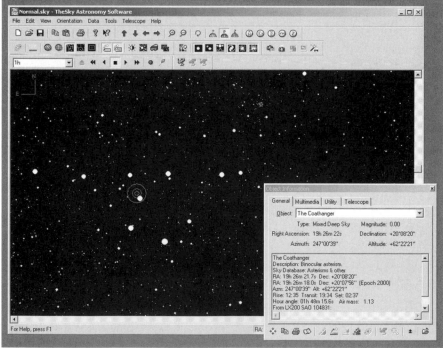

Figure 7.4. The Coathanger Cluster depicted by *Starry Night Pro* (*top*) and *TheSky* Level IV (*bottom*).

What's more, these programs can control a computerized telescope directly. When connected to the telescope, the map shows where the telescope is pointed; click on an object on the screen, and the telescope slews to it automatically. Magic!

Competing software packages are evolving rapidly, and it would be premature for me to recommend a specific one. Instead, download a free trial version and try one of them yourself.

Of the three that I have tried, *Starry Night* has the most elaborate graphics and is probably best for beginners; it even shows Mars and Jupiter with surface features and the Great Red Spot in their actual positions.

TheSky is practically a research tool for professional astronomers; it provides instant access to several catalogues and makes it easy to add your own. It is part of a suite of software comprising *CCDSOFT* (CCD camera control) and *TPoint* (precise telescope pointing); all three packages work together.

SkyMap is often considered the handiest for actual use at the telescope, especially when running on a smaller, slower computer.

Each of these programs is sold in multiple versions or "levels;" the more advanced versions have larger databases, more versatility, and higher prices. *Starry Night* and *TheSky* are available in Macintosh as well as Windows versions. For Linux there is *Xephem* (www.clearskyinstitute.com).

Star charting software is best thought of not as maps stored in a computer, but rather as the means for viewing a star catalogue visually. Normally, the Hipparcos catalogue is used for brighter stars and the Tycho and Hubble Guide Star catalogues for fainter ones. By clicking on any star or other object, you can see detailed information about it. Whether this information should be *believed* is a question that requires critical thought, since the level of precision is seldom indicated. (See for example the note about Hipparcos star distances on p. 116.) Still, it's far better than slogging through a printed book or text file to look something up.

Computer-generated plots are not as good as hand-plotted maps; despite the computer's best efforts, objects sometimes overlap and captions sometimes collide. However, the quality of computerized star maps is rising rapidly, the convenience is undeniable, and the ability to plot the orbits of the planets – including downloaded orbits of newly discovered asteroids and comets – is indispensable. My copy of *TheSky* even includes my measurements of the horizon line around my telescope pier, so that I can tell instantly whether a star is behind a neighbor's chimney.

7.3.6 Palomar Observatory Sky Survey

The ultimate star maps are the photographs comprising the *Palomar Observatory Sky Survey* (POSS), which cover the entire sky down to magnitude 20, with two images of each area, one in red light and the other in blue light.

The pictures were taken on special photographic plates with the 48-inch (1.2-meter) Schmidt camera on Palomar Mountain and later extended to the southern sky with the UK Schmidt camera in Siding Spring, Australia. For a long time they were available – as photographic prints – only in the libraries of large observatories.

Today, the entire Sky Survey has been digitized and is available online (http://archive.stsci.edu/dss) and on a set of 102 CD-ROMs. What's more, the entire sky is being rephotographed, producing a second-generation survey; it, too, is available online.

These are the definitive charts for confirming or unconfirming the existence of deep-sky objects. They have only one limitation. Because photographs tend to make stars of different brightnesses look alike, some large, loose open clusters fail to show up well on these plates and have been erroneously reported to be nonexistent.

7.4 Star catalogues

7.4.1 Online libraries

A **star catalogue** is a list of stars with exact positions and other data. Until recently, star catalogues were books, often big, thick ones. Nowadays, however, many of catalogues are published only in digital form; why print a book just so that people can look up the numbers and type them back into computers?

The central repository for catalogues of stars and other celestial objects is available online from NASA's **Astronomical Data Center** (**ADC**) at http://adc.gsfc. nasa.gov, from which catalogues can be downloaded free of charge. The format of each catalogue is documented, and software packages such as *TheSky* can accept the catalogues directly for incorporation into star maps.

If you want to look up a particular object rather than download a whole catalogue, the place to go is the **SIMBAD** astronomical database at http://simbad. u-strasbg.fr and http://simbad.harvard.edu. If you specify an object or a position, SIMBAD will retrieve all available catalogue listings, plus bibliographies of relevant astronomical research literature, and will even plot a telescopic-scale star chart. SIMBAD provides a handy way to look up an unusual designation to see if you know the same object by a more familiar name.

The name SIMBAD stands for Set of Identifications, Measurements, and Bibliography for Astronomical Data. The project is hosted by the University of Strasbourg, France.

For the online *General Catalogue of Variable Stars* see p. 138.

7.4.2 SAO

Many computerized telescopes identify stars by **SAO** numbers. These are from the *Smithsonian Astrophysical Observatory Star Catalog* (1966, 1971), which lists stars down to magnitude 9.5. Stars are numbered in a series of zones from north to south, in order of right ascension within each zone.

Only a subset of the SAO catalogue is built into any particular telescope; for example, the Meade LX200 recognizes SAO numbers for stars to magnitude 7.

7.4.3 Other bright star catalogues

The most authoritative catalogue of relatively bright stars is the Yale *Bright Star Catalogue,* by Dorrit Hoffleit (Yale University Observatory, 1982), which is usually called **YBS** although its numbering sequence is denoted **HR** (for "Harvard Revised Photometry"). An updated edition is available online from the Astronomical Data Center (http://adc.gsfc.nasa.gov).

Other classic catalogues are the *Henry Draper Catalogue* (Harvard, 1924), denoted **HD**; the *General Catalogue* of the Carnegie Institution (Washington, 1936), denoted **GC**; and the classic *Bonner Durchmusterung* [pronounced *BON-ner doorkh-MOOS-ter-oong*] of F. W. A. Argelander (1862), denoted **BD**, with its southern extension the *Córdoba Durchmusterung* (**CoD**, 1932).

SAO and GC numbers of stars to magnitude 8, along with other data, are given in *Sky Catalogue 2000.0*, vol. 1 (Cambridge University Press, 1985).

7.4.4 Hubble Guide Star Catalog

Until 1990, there was no good catalogue of stars fainter than about 10th magnitude, i.e., no catalogue of more than about 300 000 stars. The Hubble Space Telescope needed a much larger set of alignment stars, so the *Guide Star Catalog* (**GSC**) was prepared in some haste by using a computer to scan photographic plates from the Palomar Observatory Sky Survey. The resulting catalogue covers 19 million stars to about 15th magnitude.

While quite satisfactory for its purpose – aligning the Space Telescope – this catalogue is in other respects very quirky. There is a missing patch in Hercules at about $17^h20^m +33°13'$, and the magnitude limit varies from place to place across the sky. Because it is based on photographic plates, it tends to miss faint objects near bright ones; however, a number of dust specks and plate flaws appear as spurious stars. Magnitudes are to be taken with a grain of salt; they are mostly visual (white-light) in the northern sky and photographic (blue-light) in the southern sky.

A more thorough digitization of the same plates, reaching all the way to magnitude 23, has yielded the truly gigantic **USNO** (U.S. Naval Observatory) catalogue of over 520 000 000 stars. For information see http://ftp.nofs.navy.mil. A subset, with only fifty million stars, is available as a downloadable file.

7.4.5 Hipparcos and Tycho

The Hipparcos (**HIP**) catalogue contains high-quality data on stars to magnitude 12.4, complete to about 9th magnitude, measured by the European Space Agency's Hipparcos satellite. These data are much more accurate than those in the GSC. They include precise magnitude (with light curve if variable), spectral class, and parallax (to determine distance). Double stars within reach of small telescopes are, in general, treated as separate stars.

This catalogue is in fact the most accurate set of stellar data thus far made available to astronomers, although work is continuing on processing the data, and it has one notorious omission: some editions omit ξ Ursae Majoris because of difficulties measuring its proper motion (it is a double in a rapid orbit). For years to come, many astronomers' observational work will consist of examining data already gathered by Hipparcos rather than looking at the sky with telescopes.

The name *Hipparcos* is an acronym for *hi*gh *p*recision *pa*rallax-*c*orrecting *s*atellite and *almost* matches the name of the ancient Greek astronomer Hipparchos (in Latin, Hipparchus).[2] For more information see http://astro.estec.esa.nl/Hipparcos. HIP must not be confused with **HIC**, the Hipparcos Input Catalogue, a compilation of pre-existing data put together prior to the Hipparcos mission.

The Tycho catalogue (**TYC**) is a less detailed catalogue reaching magnitude 11.5 produced by another instrument on the Hipparcos satellite. It is named in honor of Tycho Brahe [pronounced *TEE-ko BRA-uh* or *BRA-hay*], who catalogued star positions in the sixteenth century, and is published along with the Hipparcos catalogue. A revised Tycho catalogue, **TYC2**, was released in 2000.

7.4.6 The cross-indexing problem

Why does each star catalogue have its own numbering system? Largely because perfect cross-indexing is impossible. What do you do when your catalogue shows two stars where another catalogue shows only one, and you want to use the other catalogue's numbering system? Your only option is to make a mistake. To avoid this, SAO, GSC, HIP, and TYC have separate numbering systems (although HIP and TYC are well cross-indexed).

I once looked up α Centauri (which is a wide double star) in a popular star-mapping software package and was startled to see five bright stars instead of the expected two. Apparently, there were two star positions from SAO or a similar conventional catalogue, two more from Hipparcos (not recognized as the same star because proper motion had taken its toll), and one from the GSC, which did not split the double.

7.4.7 Bayer/Flamsteed to SAO cross-index

With those cautionary notes I present this cross-index of Bayer and Flamsteed designations to SAO numbers. Most computerized telescopes use SAO numbers to identify stars, so if you want to observe (say) β Lyrae, you need its SAO number.

The following list gives the SAO numbers for 1120 stars down to fifth magnitude that are identified by Bayer, Flamsteed, or variable-star designations. It is based on a larger index prepared by W. Barry Smith and distributed by the Astronomical Data Center (http://adc.gsfc.nasa.gov).

---

[2] I'm probably not the only person who feels that if they could get that close, they could have found a second H somewhere!

SAO			SAO			SAO		
73765	α	And	125159	η	Aql	228069	α	Ara
54471	β	And	144150	θ	Aql	244725	β	Ara
37734	γ^1	And	143600	κ	Aql	253945	δ	Ara
54058	δ	And	143021	λ	Aql	244331	ϵ^1	Ara
74164	ϵ	And	124799	μ	Aql	244315	ζ	Ara
74267	ζ	And	105878	ρ	Aql	244168	η	Ara
53216	ι	And	125403	τ	Aql	245242	θ	Ara
53264	κ	And	104691	ω	Aql	244981	μ	Ara
53204	λ	And	143134	20	Aql	75151	α	Ari
54281	μ	And	104807	31	Aql	75012	β	Ari
36699	ν	And	143482	36	Aql	93328	δ	Ari
37155	ξ	And	163036	51	Aql	75810	ζ	Ari
52609	o	And	144495	69	Aql	92877	θ	Ari
54033	π	And	145862	α	Aqr	75495	ν	Ari
53828	ρ	And	145457	β	Aqr	93144	σ	Ari
53798	σ	And	146044	γ	Aqr	75886	τ	Ari
37362	υ	And	165375	δ	Aqr	92637	4	Ari
53355	ψ	And	144810	ϵ	Aqr	92822	15	Ari
37228	ω	And	146181	η	Aqr	92841	19	Ari
52713	5	And	145991	θ	Aqr	75238	21	Ari
73190	12	And	164861	ι	Aqr	92983	27	Ari
73346	15	And	146210	κ	Aqr	75532	35	Ari
36123	22	And	146362	λ	Aqr	75596	41	Ari
37375	51	And	144895	μ	Aqr	75662	47	Ari
37948	62	And	164182	ν	Aqr	75757	55	Ari
201405	α	Ant	145537	ξ	Aqr	40186	α	Aur
200416	ϵ	Ant	127520	π	Aqr	40750	β	Aur
200926	η	Ant	165134	σ	Aqr	25502	δ	Aur
177908	θ	Ant	165321	τ	Aqr	39955	ϵ	Aur
201927	ι	Ant	191235	υ	Aqr	39966	ζ	Aur
257193	α	Aps	146585	ϕ	Aqr	40026	η	Aur
257407	γ	Aps	146598	ψ^1	Aqr	57522	ι	Aur
257380	δ^1	Aps	146635	ψ^3	Aqr	78143	κ	Aur
257112	θ	Aps	165842	ω^2	Aqr	40233	λ	Aur
257491	ι	Aps	144814	3	Aqr	57755	μ	Aur
257289	κ^1	Aps	145022	11	Aqr	58502	ν	Aur
125122	α	Aql	164364	18	Aqr	25450	ξ	Aur
125235	β	Aql	191083	47	Aqr	40583	o	Aur
105223	γ	Aql	165293	68	Aqr	58164	χ	Aur
124603	δ	Aql	191683	88	Aqr	41076	ψ^1	Aur
104318	ϵ	Aql	191858	98	Aqr	41330	ψ^5	Aur
104461	ζ	Aql	165854	106	Aqr	41346	ψ^6	Aur

SAO			SAO			SAO		
59316	51	Aur	163422	α^1	Cap	252582	β	Cen
59866	63	Aur	163427	α^2	Cap	239689	δ	Cen
41679	64	Aur	163481	β	Cap	241047	ϵ	Cen
41738	66	Aur	164560	γ	Cap	224538	ζ	Cen
100944	α	Boo	164644	δ	Cap	225044	η	Cen
45337	β	Boo	190341	ζ	Cap	205188	θ	Cen
64203	γ	Boo	164132	θ	Cap	204371	ι	Cen
64589	δ	Boo	164346	ι	Cap	225344	κ	Cen
100766	η	Boo	164639	λ	Cap	251472	λ	Cen
29137	θ	Boo	164713	μ	Cap	224471	μ	Cen
29071	ι	Boo	163779	υ	Cap	223909	ξ^2	Cen
44965	λ	Boo	189664	ψ	Cap	238986	π	Cen
64686	μ^1	Boo	189781	ω	Cap	223454	σ	Cen
45580	ν^1	Boo	189114	4	Cap	205453	ψ	Cen
64202	ρ	Boo	190025	24	Cap	204812	1	Cen
83416	σ	Boo	234480	α	Car	19302	α	Cep
100706	τ	Boo	250495	β	Car	10057	β	Cep
45643	ϕ	Boo	235932	ϵ	Car	10818	γ	Cep
83645	ψ	Boo	251083	θ	Car	34508	δ	Cep
82993	3	Boo	236808	ι	Car	34137	ζ	Cep
83130	11	Boo	250905	q	Car	18897	θ	Cep
83203	12	Boo	235635	χ	Car	20268	ι	Cep
100975	18	Boo	250885	ω	Car	9665	κ	Cep
101025	22	Boo	21609	α	Cas	19624	ν	Cep
101152	32	Boo	21133	β	Cas	10402	ρ	Cep
45153	33	Boo	11482	γ	Cas	3994	V	Cep
83488	34	Boo	22268	δ	Cas	19019	7	Cep
83671	45	Boo	12031	ϵ	Cas	10126	11	Cep
45370	47	Boo	21566	ζ	Cas	19847	20	Cep
216926	α	Cae	11256	κ	Cas	10265	24	Cep
195239	β	Cae	22024	μ	Cas	20190	30	Cep
216850	δ	Cae	36620	o	Cas	10425	31	Cep
13298	α	Cam	35879	ρ	Cas	110920	α	Cet
13351	β	Cam	11751	ψ	Cas	147420	β	Cet
5006	γ	Cam	20614	4	Cas	110665	δ	Cet
24672	1	Cam	4216	21	Cas	148059	ζ	Cet
24829	4	Cam	4422	38	Cas	147632	η	Cet
13518	17	Cam	4453	40	Cas	129274	θ	Cet
25241	18	Cam	11919	43	Cas	128694	ι	Cet
13756	36	Cam	4560	50	Cas	111120	κ	Cet
13986	43	Cam	12180	55	Cas	110889	λ	Cet
14402	53	Cam	252838	α	Cen	110723	μ	Cet

SAO			SAO			SAO		
110635	ν	Cet	96952	6	CMi	179624	β	Crt
110408	ξ^1	Cet	97224	11	CMi	156661	γ	Crt
110543	ξ^2	Cet	98267	α	Cnc	156605	δ	Crt
148575	π	Cet	116569	β	Cnc	156869	ζ	Crt
147470	π^2	Cet	80378	γ	Cnc	156988	η	Crt
148385	ρ	Cet	98087	δ	Cnc	138296	θ	Crt
148445	σ	Cet	80243	η	Cnc	251904	α^1	Cru
147986	τ	Cet	80416	ι	Cnc	240259	β	Cru
167471	υ	Cet	98378	κ	Cnc	240019	γ	Cru
148036	χ	Cet	80666	ξ	Cnc	239791	δ	Cru
147059	2	Cet	61177	σ^3	Cnc	180915	β	Crv
128791	12	Cet	80104	χ	Cnc	157176	γ	Crv
129009	20	Cet	79861	ω	Cnc	157323	δ	Crv
109643	26	Cet	97399	1	Cnc	180531	ϵ	Crv
147812	47	Cet	97781	20	Cnc	63257	α^2	CVn
167086	48	Cet	97843	29	Cnc	44230	β	CVn
129798	67	Cet	196059	α	Col	44317	Y	CVn
130004	80	Cet	196240	β	Col	44097	2	CVn
130355	94	Cet	196352	γ	Col	44127	3	CVn
256924	β	Cha	217650	η	Col	63000	6	CVn
256731	γ	Cha	196643	κ	Col	63338	14	CVn
258593	δ^2	Cha	195721	o	Col	63380	17	CVn
256543	η	Cha	82706	β	Com	44549	20	CVn
256503	θ	Cha	99973	3	Com	44570	23	CVn
256857	π	Cha	82273	12	Com	49941	α	Cyg
252852	α	Cir	82336	20	Com	87301	β	Cyg
242384	β	Cir	82390	23	Com	49528	γ	Cyg
151881	α	CMa	100160	24	Com	70474	ϵ	Cyg
151428	β	CMa	82537	31	Com	71070	ζ	Cyg
152303	γ	CMa	100309	32	Com	69116	η	Cyg
173047	δ	CMa	82650	39	Com	31815	θ	Cyg
172676	ϵ	CMa	210990	α	CrA	31702	ι	Cyg
196698	ζ	CMa	229299	η	CrA	31537	κ	Cyg
173651	η	CMa	229111	θ	CrA	50274	ν	Cyg
152071	θ	CMa	83893	α	CrB	50424	ξ	Cyg
197258	κ	CMa	83831	β	CrB	49337	o^1	Cyg
171982	ξ^2	CMa	84098	ϵ	CrB	51293	π^2	Cyg
172839	o^2	CMa	64769	θ	CrB	51035	ρ	Cyg
172797	σ	CMa	64948	κ	CrB	71165	σ	Cyg
115756	α	CMi	65108	τ	CrB	71173	υ	Cyg
115456	β	CMi	156375	α	Crt	68447	8	Cyg
116043	ζ	CMi				68778	15	Cyg

SAO			SAO			SAO		
69518	28	Cyg	17526	27	Dra	167532	ν	For
32378	33	Cyg	8939	35	Dra	168701	τ	For
70095	41	Cyg	17828	36	Dra	60198	α	Gem
70096	42	Cyg	9250	50	Dra	79666	β	Gem
50335	59	Cyg	9802	73	Dra	95912	γ	Gem
70919	61	Cyg	3458	76	Dra	79294	δ	Gem
50934	71	Cyg	126662	α	Equ	78682	ϵ	Gem
51101	74	Cyg	126593	γ	Equ	79031	ζ	Gem
106357	α	Del	232481	α	Eri	59570	θ	Gem
106476	γ^2	Del	131794	β	Eri	79374	ι	Gem
106425	δ	Del	149283	γ	Eri	79653	κ	Gem
106230	ϵ	Del	130686	δ	Eri	96746	λ	Gem
126059	κ	Del	130564	ϵ	Eri	78297	μ	Gem
233564	α	Dor	130387	ζ	Eri	78423	ν	Gem
249311	β	Dor	130197	η	Eri	96074	ξ	Gem
233457	γ	Dor	216113	θ^1	Eri	60340	π	Gem
249346	δ	Dor	215999	ι	Eri	60118	ρ	Gem
233822	ζ	Dor	215906	κ	Eri	79533	υ	Gem
249225	θ	Dor	131824	λ	Eri	79896	χ	Gem
249461	ν	Dor	131468	μ	Eri	79774	ψ	Gem
16273	α	Dra	131346	ν	Eri	78999	ω	Gem
30429	β	Dra	131176	ξ	Eri	77915	1	Gem
30653	γ	Dra	131019	o^1	Eri	96638	51	Gem
18222	δ	Dra	168094	τ^2	Eri	97221	81	Gem
17365	ζ	Dra	168249	τ^3	Eri	230992	α	Gru
29765	θ	Dra	168634	τ^5	Eri	231258	β	Gru
29520	ι	Dra	168827	τ^6	Eri	213374	γ	Gru
7593	κ	Dra	195148	υ^2	Eri	231154	δ^1	Gru
15532	λ	Dra	232696	ϕ	Eri	247593	ϵ	Gru
30447	ν^1	Dra	232573	χ	Eri	247680	ζ	Gru
30450	ν^2	Dra	130528	17	Eri	231468	ι	Gru
30631	ξ	Dra	149063	20	Eri	213543	λ	Gru
31219	o	Dra	130698	24	Eri	213850	ν	Gru
9366	τ	Dra	130878	35	Eri	247874	o	Gru
9283	υ	Dra	194984	43	Eri	84411	β	Her
9087	χ	Dra	149781	53	Eri	102107	γ	Her
8890	ψ	Dra	131451	56	Eri	84951	δ	Her
17576	ω	Dra	193931	β	For	65716	ϵ	Her
15606	3	Dra	194467	δ	For	65504	η	Her
15941	8	Dra	167736	κ	For	66485	θ	Her
16199	10	Dra	193763	λ^1	For	46872	ι	Her
17107	15	Dra	193573	μ	For	101951	κ	Her

SAO			SAO			SAO		
85163	λ	Her	155713	υ^2	Hya	81064	μ	Leo
85397	μ	Her	179514	χ^1	Hya	98627	ξ	Leo
85590	ξ	Her	154515	6	Hya	98709	o	Leo
85750	o	Her	136308	14	Hya	118044	π	Leo
65890	π	Her	136832	28	Hya	118355	ρ	Leo
46161	σ	Her	178979	44	Hya	118804	σ	Leo
46028	τ	Her	182134	47	Hya	118875	τ	Leo
45911	ϕ	Her	182152	48	Hya	138298	υ	Leo
45772	χ	Her	182570	52	Hya	138102	ϕ	Leo
102153	ω	Her	182882	56	Hya	118648	χ	Leo
121568	21	Her	248474	α	Hyi	98733	ψ	Leo
46210	42	Her	255670	β	Hyi	61656	15	Leo
102435	49	Her	256029	γ	Hyi	99172	46	Leo
84651	51	Her	248545	δ	Hyi	99281	51	Leo
65627	53	Her	248621	ϵ	Hyi	99305	53	Leo
65736	59	Her	248460	η^2	Hyi	118610	58	Leo
102584	60	Her	255945	θ	Hyi	118668	65	Leo
65963	72	Her	255973	ι	Hyi	118864	83	Leo
46723	77	Her	255880	κ	Hyi	81998	93	Leo
85545	89	Her	255710	λ	Hyi	99869	95	Leo
103285	93	Her	255898	μ	Hyi	150547	α	Lep
86003	109	Her	230300	α	Ind	170457	β	Lep
86406	110	Her	246784	β	Ind	170759	γ	Lep
104093	111	Her	247031	γ	Ind	170926	δ	Lep
216710	α	Hor	247244	δ	Ind	170051	ϵ	Lep
232857	ζ	Hor	247287	ϵ	Ind	150801	ζ	Lep
248555	λ	Hor	246709	η	Ind	150957	η	Lep
232981	μ	Hor	255087	o	Ind	150340	λ	Lep
136871	α	Hya	258084	ρ	Ind	150237	μ	Lep
181543	γ	Hya	34542	α	Lac	158836	α^1	Lib
116965	δ	Hya	34395	β	Lac	158840	α^2	Lib
117264	ζ	Hya	72575	10	Lac	140430	β	Lib
117527	θ	Hya	52317	13	Lac	159370	γ	Lib
137035	ι	Hya	98967	α	Leo	140270	δ	Lib
155388	κ	Hya	99809	β	Leo	159090	ι	Lib
155785	λ	Hya	81727	δ	Leo	159442	κ	Lib
155980	μ	Hya	81004	ϵ	Leo	183895	λ	Lib
156256	ν	Hya	81265	ζ	Leo	158915	ξ^2	Lib
202558	ξ	Hya	98955	η	Leo	183139	σ	Lib
202695	o	Hya	99512	θ	Leo	183619	υ	Lib
182244	π	Hya	80807	κ	Leo	158528	2	Lib
116988	σ	Hya				159280	32	Lib

SAO			SAO			SAO		
140609	37	Lib	256201	γ	Men	258932	υ	Oct
159607	48	Lib	258372	δ	Men	258799	χ	Oct
159625	49	Lib	258451	ζ	Men			
140897	50	Lib	256145	η	Men	102932	α	Oph
62053	β	LMi	256122	μ	Men	122671	β	Oph
61570	10	LMi	258395	ξ	Men	122754	γ	Oph
43115	19	LMi	212636	γ	Mic	141052	δ	Oph
61808	20	LMi	212874	ϵ	Mic	141086	ϵ	Oph
62010	27	LMi	212666	ζ	Mic	160006	ζ	Oph
62173	37	LMi	230644	θ^1	Mic	185320	θ	Oph
81490	41	LMi	230379	ι	Mic	102458	ι	Oph
62236	42	LMi	134986	α	Mon	121962	κ	Oph
62297	46	LMi	134330	δ	Mon	142004	ν	Oph
225128	α	Lup	133114	7	Mon	122387	σ	Oph
225335	β	Lup	113810	8	Mon	122226	U	Oph
225691	δ	Lup	133290	10	Mon	141269	12	Oph
242304	ζ	Lup	114034	13	Mon	121859	19	Oph
207332	θ	Lup	114388	16	Mon	160118	20	Oph
225525	κ	Lup	114428	18	Mon	141483	30	Oph
224919	τ^1	Lup	134282	20	Mon	185401	44	Oph
206552	ϕ^1	Lup	134899	25	Mon	185412	45	Oph
206580	ϕ^2	Lup	135345	27	Mon	185660	58	Oph
207040	χ	Lup	251974	α	Mus	123013	67	Oph
206445	1	Lup	256955	γ	Mus	123142	72	Oph
61414	α	Lyn	257000	δ	Mus	123377	74	Oph
25665	2	Lyn	252224	η	Mus	113271	α	Ori
13897	8	Lyn	251575	λ	Mus	131907	β	Ori
26312	19	Lyn	243643	γ^2	Nor	112740	γ	Ori
26474	24	Lyn	226500	δ	Nor	132220	δ	Ori
42058	26	Lyn	243454	κ	Nor	132346	ϵ	Ori
26687	27	Lyn	257879	α	Oct	132323	ι	Ori
42319	31	Lyn	258941	β	Oct	132542	κ	Ori
42490	34	Lyn	258928	ϵ	Oct	95259	ν	Ori
42759	36	Lyn	258515	ζ	Oct	94176	o^1	Ori
67174	α	Lyr	258600	η	Oct	112106	π^3	Ori
67451	β	Lyr	258207	θ	Oct	112142	π^4	Ori
67663	γ	Lyr	258654	ι	Oct	112197	π^5	Ori
68065	θ	Lyr	258674	κ	Oct	131952	τ	Ori
67834	ι	Lyr	257948	ν	Oct	112914	ϕ^1	Ori
66869	κ	Lyr	258731	ρ	Oct	94290	11	Ori
47919	13	Lyr	258857	σ	Oct	112467	16	Ori
256274	α	Men	258970	τ	Oct	132028	22	Ori
						113321	60	Ori

SAO			SAO			SAO		
113430	66	Ori	73241	67	Peg	213883	β	PsA
95476	74	Ori	108638	70	Peg	191318	ϵ	PsA
			108879	82	Peg	213258	ι	PsA
246574	α	Pav	38787	α	Per	190985	λ	PsA
254862	β	Pav	38592	β	Per	213576	μ	PsA
254999	γ	Pav	23789	γ	Per	214275	π	PsA
254733	δ	Pav	39053	δ	Per	213078	6	PsA
257757	ϵ	Pav	56840	ϵ	Per			
257620	ζ	Pav	56799	ζ	Per	127934	β	Psc
254020	η	Pav	23655	η	Per	128085	γ	Psc
254393	λ	Pav	38288	θ	Per	109474	δ	Psc
254226	ξ	Pav	38597	ι	Per	109627	ϵ	Psc
257896	o	Pav	24412	λ	Per	109739	ζ	Psc
			39404	μ	Per	92484	η	Psc
108378	α	Peg	39078	ν	Per	128196	θ	Psc
90981	β	Peg	56856	ξ	Per	128310	ι	Psc
91781	γ	Peg	56138	ρ	Per	128186	κ	Psc
127029	ϵ	Peg	38890	σ	Per	128336	λ	Psc
108103	ζ	Peg	23685	τ	Per	110065	ν	Psc
90734	η	Peg	22554	ϕ	Per	110206	ξ	Psc
127340	θ	Peg	22696	2	Per	110110	o	Psc
90238	ι	Peg	22859	4	Per	92536	π	Psc
90775	λ	Peg	38289	14	Per	74546	τ	Psc
90816	μ	Peg	56052	24	Per	74637	υ	Psc
72077	π	Peg	39336	48	Per	74544	χ	Psc
91186	τ	Peg	57171	54	Per	128513	ω	Psc
91253	υ	Peg	39604	57	Per	146915	20	Psc
108878	ϕ	Peg				147008	27	Psc
91792	χ	Peg	215093	α	Phe	147042	30	Psc
91611	ψ	Peg	215516	γ	Phe	128572	33	Psc
107073	1	Peg	214983	ϵ	Phe	109152	41	Psc
89752	2	Peg	232162	η	Phe	109192	44	Psc
107288	5	Peg	231675	ι	Phe	91912	48	Psc
127060	11	Peg	215131	λ^1	Phe	92099	64	Psc
90040	14	Peg	215194	o	Phe	74395	68	Psc
90075	16	Peg	248087	π	Phe	92230	72	Psc
107587	20	Peg	215696	ψ	Phe	109793	89	Psc
72064	27	Peg				92444	94	Psc
107854	31	Peg	249647	α	Pic			
127544	36	Peg	234154	γ	Pic	198752	ζ	Pup
72406	38	Peg	234359	δ	Pic	218071	ν	Pup
108154	45	Peg	233926	ζ	Pic	174601	ξ	Pup
127976	55	Peg	217164	η^2	Pic	197795	π	Pup
128022	59	Peg	191524	α	PsA	175217	ρ	Pup

SAO			SAO			SAO		
218755	σ	Pup	142515	δ	Sct	162518	υ	Sgr
234735	τ	Pup	142546	ϵ	Sct	187239	ϕ	Sgr
153372	4	Pup	121157	α	Ser	185755	3	Sgr
153993	20	Pup	101725	β	Ser	160998	6	Sgr
199546	α	Pyx	101826	γ	Ser	187342	30	Sgr
176559	γ	Pyx	121218	ϵ	Ser	162413	43	Sgr
200047	ϵ	Pyx	142241	η	Ser	188337	52	Sgr
177322	θ	Pyx	124068	θ^1	Ser	162883	54	Sgr
248969	α	Ret	101752	κ	Ser	162915	55	Sgr
248877	β	Ret	140787	μ	Ser	162964	56	Sgr
248918	δ	Ret	160700	ξ	Ser	163141	61	Sgr
249009	η	Ret	121540	σ	Ser	188844	62	Sgr
248819	κ	Ret	101545	τ^1	Ser	94027	α	Tau
166716	α	Scl	120916	3	Ser	77168	β	Tau
214615	β	Scl	140502	8	Ser	93868	γ	Tau
214444	γ	Scl	142348	60	Ser	93897	δ^1	Tau
192167	δ	Scl	137600	δ	Sex	93954	ϵ	Tau
167275	ϵ	Scl	137469	ϵ	Sex	77336	ζ	Tau
192388	θ	Scl	137183	6	Sex	76199	η	Tau
166103	κ^2	Scl	118041	12	Sex	76920	ι	Tau
192703	λ^2	Scl	118248	23	Sex	93719	λ	Tau
214701	μ	Scl	137533	25	Sex	111696	μ	Tau
193263	π	Scl	137728	33	Sex	111579	ν	Tau
192884	σ	Scl	137823	41	Sex	111195	ξ	Tau
184415	α	Sco	105133	β	Sge	111172	o	Tau
159682	β^1	Sco	105500	γ	Sge	94007	ρ	Tau
184014	δ	Sco	105259	δ	Sge	76721	τ	Tau
208078	ϵ	Sco	229659	α	Sgr	93469	5	Tau
227707	η	Sco	229646	β^1	Sgr	111292	10	Tau
228201	θ	Sco	209696	γ	Sgr	76073	11	Tau
228420	ι^1	Sco	186681	δ	Sgr	76131	17	Tau
209163	κ	Sco	210091	ϵ	Sgr	76228	27	Tau
208954	λ	Sco	209957	η	Sgr	111400	29	Tau
208102	μ^1	Sco	211716	θ^1	Sgr	76430	37	Tau
183987	π	Sco	229927	ι	Sgr	93785	43	Tau
184336	σ	Sco	230177	κ^1	Sgr	76485	44	Tau
184481	τ	Sco	186841	λ	Sgr	94164	97	Tau
208896	υ	Sco	186497	μ	Sgr	94554	115	Tau
209318	G	Sco	187504	ξ^2	Sgr	94858	130	Tau
142408	α	Sct	187756	π	Sgr	77675	136	Tau
142618	β	Sct	187448	σ	Sgr	229023	α	Tel
161520	γ	Sct	187683	τ	Sgr	228777	ϵ	Tel

SAO				SAO				SAO		
229751	ι	Tel		43886	χ	UMa		158427	κ	Vir
245834	λ	Tel		43629	ψ	UMa		158489	λ	Vir
246271	ν	Tel		14908	23	UMa		140090	μ	Vir
246443	ξ	Tel		6897	24	UMa		119035	ν	Vir
253700	α	TrA		15135	32	UMa		119213	o	Vir
253346	β	TrA		27670	36	UMa		119164	π	Vir
253097	γ	TrA		27695	37	UMa		100211	ρ	Vir
253474	δ	TrA		43557	47	UMa		119855	σ	Vir
253226	ϵ	TrA		62491	55	UMa		120238	τ	Vir
253554	ζ	TrA		43787	58	UMa		139951	ϕ	Vir
				62655	61	UMa		138892	χ	Vir
74996	α	Tri		28405	74	UMa		139033	ψ	Vir
55306	β	Tri		15871	76	UMa		119341	16	Vir
55427	γ	Tri						138873	25	Vir
75382	12	Tri		308	α	UMi		119574	32	Vir
55635	14	Tri		8102	β	UMi		119596	35	Vir
				8220	γ	UMi		139086	44	Vir
255193	α	Tuc		2937	δ	UMi		157844	61	Vir
247814	γ	Tuc		2770	ϵ	UMi		157938	68	Vir
255619	ϵ	Tuc		8328	ζ	UMi		100582	70	Vir
248163	ζ	Tuc		8470	η	UMi		120004	78	Vir
248324	ι	Tuc		3020	λ	UMi		139428	80	Vir
248281	λ^2	Tuc		16558	RR	UMi		139490	82	Vir
				7958	4	UMi		158131	83	Vir
15384	α	UMa		8024	5	UMi		158186	89	Vir
27876	β	UMa		8446	19	UMi		120648	109	Vir
28179	γ	UMa								
28315	δ	UMa		219504	γ	Vel		250422	α	Vol
28553	ϵ	UMa		236891	κ	Vel		250228	β	Vol
28737	ζ	UMa		220878	λ	Vel		256374	γ^2	Vol
44752	η	UMa		236164	o	Vel		249809	δ	Vol
27289	θ	UMa		237522	ϕ	Vel		256438	ζ	Vol
42630	ι	UMa						256344	ι	Vol
42661	κ	UMa		157923	α	Vir				
43268	λ	UMa		119076	β	Vir		87261	α	Vul
43310	μ	UMa		119674	δ	Vir		87633	10	Vul
62486	ν	UMa		100384	ϵ	Vir		88071	15	Vul
62484	ξ	UMa		139420	ζ	Vir		88451	24	Vul
14573	o	UMa		138721	η	Vir		88944	29	Vul
14742	ρ	UMa		139189	θ	Vir		89272	32	Vul
27401	υ	UMa		139824	ι	Vir		89332	33	Vul

Chapter 8
Stars – physical properties

8.1 Magnitude

8.1.1 The magnitude system

The **magnitude** of a star is its brightness measured in a somewhat peculiar way. In ancient times, Ptolemy and Hipparchos classified stars as "first class" (brightest) to "sixth class" (barely visible). These brightness classes were termed *magnitudes,* but there was no provision for exact measurement.

In 1856, Norman Pogson proposed the logarithmic magnitude scale that is now standard. The advantage of a logarithmic scale is that it can span a tremendous brightness range without using very large or very small numbers (Figure 8.1). Each difference of five magnitudes corresponds to a factor-of-100 difference in brightness. One magnitude corresponds to a brightness ratio of 2.512.

In this system, most stars still have roughly the magnitude that Ptolemy assigned them, but some of the brightest stars have negative magnitudes. The full moon is magnitude −12 and the Sun is −27. This 15-magnitude difference means that the Sun is a million times as bright as the Moon.

The star Vega is defined to be magnitude 0.0, but in practice, the average of several stars is used as a standard for measurement.

The human eye's response to light is not actually logarithmic, but it is close enough for practical purposes. If a star appears to be halfway between two other stars in brightness, it will also be halfway between them in magnitude.

On *absolute magnitude* see p. 117.

8.1.2 Calculations with magnitudes

Because magnitudes are logarithmic, you cannot add or subtract them directly. You can, however, convert magnitude (m) into brightness measured on a linear scale (L), perform the addition or subtraction, and then convert the result back into a magnitude.

Relative brightness	Magnitude	Example
1 000 000	**−15**	−12 full moon
10,000	**−10**	
100	**−5**	−4.3 Venus
1	**0**	−1.4 Sirius 0.0 Vega 2.0 Polaris
1/100	**5**	6.0 Naked-eye limit
1 / 10 000	**10**	10.0 Limit of 7 x 35 binoculars
1 / 1 000 000	**15**	14.0 Limit of 8-inch telescope

Figure 8.1. The magnitude scale is logarithmic; every five magnitude steps correspond to a brightness ratio of 100:1.

Here are the formulae:

$$L = 10^{-0.4m}$$

$$m = -2.5 \log_{10} L$$

The number 2.5 in the second formula is exact; it is not an approximation for 2.512. Note that on many calculators, pressing INV LOG computes 10^x.

Example: The double star Σ 2398 has components of $m = 8.9$ and 9.7. What is the combined magnitude when the two stars are not separated?

Convert to L:

$$L_1 = 10^{-0.4 \times 8.9} = 0.000\,275$$

$$L_2 = 10^{-0.4 \times 9.7} = 0.000\,132$$

Add:

$$0.000\,275 + 0.000\,132 = 0.000\,407$$

Convert back to m:

$$m = -2.5 \log_{10} 0.000\,407 = 8.5$$

Thus the two stars together look like a single star of magnitude 8.5. As a rule, two stars, combined, will never be more than 0.25 magnitude brighter than the brighter star in the pair.

8.1.3 Telescope magnitude limits

The magnitude limit of the human eye is about 6 (or 7 at really dark, clear desert and mountain sites). With a telescope, you can see stars down to much fainter magnitudes. In theory, the magnitude limit depends only on the telescope

Table 8.1. *Expected magnitude limit vs. telescope size*

Aperture		Magnitude
inches	cm	limit
2.4	6	11.3
3.5	9	12.3
5	12.5	13.0
8	20	14.0
12	30	14.9

aperture, not the magnification, because stars form point images. In practice, medium to high power shows fainter stars than low power.

Here are formulae to find the approximate magnitude limit of a telescope of a given size with a moderately experienced observer under average conditions. A skilled observer under dark skies can do as much as two magnitudes better.

$$\text{Limiting magnitude} = 7.5 + 5 \log_{10} \text{aperture (cm)}$$

$$\text{Limiting magnitude} = 9.5 + 5 \log_{10} \text{aperture (inches)}$$

Table 8.1 shows the magnitude limits for several popular sizes of telescopes.

A central obstruction of 40% of the telescope aperture (as in some Schmidt–Cassegrains) costs you only 0.2 magnitude because it blocks only 16% (= 40%²) of the light, and $-2.5 \log_{10}(1 - 0.16) \approx 0.2$.

8.1.4 Magnitudes in old books

Before Pogson standardized the magnitude system, different telescopic observers extended the system in different ways. Magnitude 16 in Smyth's *A Cycle of Celestial Objects* corresponds to 13 in the *Bonner Durchmusterung* (which is close to the modern system) and a mere 10.9 in the double-star lists of F. G. W. Struve. See T. W. Webb, *Celestial Objects for Common Telescopes* (1917, repr. Dover, 1962), vol. 2, p. 7.

8.2 Number of stars in the sky

Any time you increase your magnitude limit by 1, you see three times as many stars in the same area of sky. Thus, there are about 2000 stars in the whole sky brighter than magnitude 5.0, about 6000 brighter than 6.0, and so on.

This is an empirical factor. If stars were uniformly distributed and there were no interstellar dust, the factor would be not 3 but 3.98.

Astrophotographers looking for guide stars often need to know how many stars they can expect to see in a field of a certain size. Table 8.2 shows the number

Table 8.2. *Number of stars in the sky*

Magnitude limit	Whole sky	In 1° square field		
		Average field	Rich field	Sparse field
1.5	22			
2.5	93			
3.5	283			
4.5	893			
5.5	2822			
6.5	8768			
7.5	26 533			
8.5	77 627	2	4	1
9.5	217 689	5	10	3
10.5	≈600 000	15	29	7
11.5	≈1 800 000	44	87	22
12.5	≈5 400 000	130	260	65

of stars above various magnitude limits in the whole sky and in a field 1° square. Values up to $m = 9.5$ are exact (from the *Millennium Star Atlas*); those for higher magnitudes are estimated.

The area of the whole celestial sphere is 4π steradians (square radians) = 41 252.9 square degrees.

8.3 Distances of the stars

How far away are the stars? The answer is complicated and involves many methods of measuring distance, each of them used as a basis for calibrating the next. It's easy to look up someone's guess as to the distance of a celestial object, but it's not always obvious whether the guess is likely to be accurate.

8.3.1 Distance units

One **astronomical unit** (AU) is the mean distance from the Earth to the Sun, about 93 000 000 miles or 150 000 000 kilometers. The AU is the basis of the stellar parallax scale.

One **light-year** (**ly**) is the distance light travels in a year, approximately 5 800 000 000 000 miles or 9 500 000 000 000 kilometers. When you look at the Andromeda Galaxy, 2 million light-years away, you are seeing light that left it 2 million years ago. The nearest star other than the Sun is Proxima Centauri (α Centauri C) at 4.3 ly. The nearest star that you can see from Europe or the United States is Barnard's Star, at 6 ly.

One **parsec** (**pc**) is the distance of a star whose parallax, over a distance of 1 AU, would be 1″. One parsec is about 3.26 light-years. **Kiloparsecs** (= 1000 pc) and **megaparsecs** (= 1 000 000 pc) are also used.

8.3.2 Parallax

Whenever you move your head, you experience **parallax.** Nearby objects seem to shift relative to more distant objects as your point of view changes. Parallax is the most reliable way of measuring celestial distances.

Distances within the Solar System are measured by means of parallax between different points on the Earth's surface. For example, if an observer in England and one in California look at the Moon or a planet at the same time, they will see it in slightly different positions against the background of stars. If the distance between the observers is known, the distance of the Moon or planet can be calculated.

Once one distance is known, numerous other distances can be deduced from it because they are related by known orbits. Measurements of this kind – including occultation timings made by amateurs – give increasingly precise values for the Earth's mean distance from the Sun.

Distances to the nearest stars are then measured by parallax from one side of the Earth's orbit to the other. A nearby star, observed in June and again in December, will seem to shift relative to distant stars in the background. The shift is tiny, always less than 1″.

This is what is called the **parallax of a star**, always quoted relative to a baseline of 1 AU. The Hipparcos satellite recently measured star parallaxes to a precision of 0.001″ (1 milliarcsecond or **mas**), ten times better than the best Earth-based measurements. Distances measured this way are accurate to within 10% at distances up to 300 light-years.

Beyond that, distances measured by parallax are not necessarily as precise as they look. For instance, if a Hipparcos-derived database gives the distance of a star as "658 light-years," what that means is that the parallax was measured as 5 mas, barely distinguishable from 4 mas (823 light-years) or 6 mas (548 light-years).

8.3.3 Measuring greater distances

When a star is beyond the reach of parallax, its distance cannot be measured precisely unless it is part of a cluster or galaxy that can be studied as a unit. One of the most important techniques is the **moving cluster method.** This involves finding a group of stars, typically an open cluster, whose proper motions seem to be converging to a point. In reality, the stars are moving in parallel, and the angle from which we are viewing their motion can be determined. This, together with the radial velocity (measured spectroscopically), gives the distance.

Another tactic is to deduce the true brightness of a star from its spectrum or other characteristics, then compare the true brightness to the apparent brightness and estimate the distance. This works particularly well with Cepheid variable stars (p. 133), whose period correlates very well with luminosity. The Cepheid distance scale gives us a way to determine the distance of M31 and other nearby galaxies in which individual stars can be observed.

If a galaxy isn't close enough to show Cepheids, it may nonetheless have supergiant stars and hydrogen nebulae that we can see from Earth. The distance can be estimated by assuming that these objects are comparable to similar objects in our own galaxy and its neighbors.

Finally, the distances of the most remote galaxies are estimated from their redshift (radial velocity), which is proportional to distance (**Hubble's Law**).

8.3.4 Absolute magnitude

The **absolute magnitude** of a star, M, is the apparent magnitude that it would have at a distance of 10 parsecs. The absolute magnitude of a planet or asteroid is the magnitude that it would have if it were 1 AU from the Earth and also 1 AU from the Sun.

Absolute magnitudes of stars range from about -5 (for the largest supergiants) to 15 (for the smallest observable dwarfs). The absolute magnitude of the Sun, an average-sized star, is 4.8.

Apparent and absolute magnitudes are related by the formulae:

$$M = m - 5\log_{10}\frac{\text{distance in pc}}{10} = m - 5\log_{10}\frac{\text{distance in ly}}{32.6}$$

$$\text{Distance (parsecs)} = 10^x \quad \text{where} \quad x = \frac{m-M}{5} + 1$$

For example, Sirius has apparent magnitude $m = -1.4$, and its distance is 2.64 parsecs. The first formula tells us that its absolute magnitude is 1.5, considerably brighter than the Sun.

8.4 Colors and spectra

8.4.1 Star colors

In the nineteenth century, observers such as Smyth and Webb recorded many fanciful colors for double stars, such as "greenish and lilac." Some of these colors have a physical basis; others probably indicate uncorrected chromatic aberration in old refracting telescopes. Any star can be endowed with a purple fringe and/or reddish center by viewing it slightly out of focus in a large refractor with a two-element (non-apochromatic) lens.

Stars do range from reddish to blue-white, but most "red" stars are no redder than an ordinary light bulb (Figure 8.2). Lower-temperature stars are redder, of course, but they are also fainter, and we can't see very many of them.

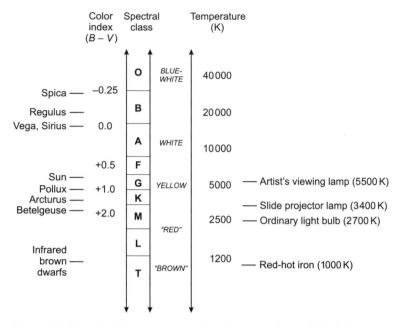

Figure 8.2. Most "red" stars are no redder than an ordinary light bulb. Correlation between color index, spectral class, and temperature is not perfect.

Besides low temperature, a star can be reddened by interstellar dust (which transmits more red light than blue) and by carbon in its composition.

8.4.2 *B* and *V* magnitudes; color index

The magnitude of a star can be measured at any wavelength of light. Standard wavelength bands for photometry are:

U   (ultraviolet)   300–400 nm
B   (blue)   360–550 nm (440 nm peak)
V   (visual)   480–680 nm (550 nm peak)
R   (red)   530–950 nm
I   (infrared)   700–1200 nm

The most important are B and V, which correspond to the response of old blue-sensitive photographic plates and the human eye respectively.

The star Vega is defined to be magnitude 0.0 at all wavelengths. This does not mean that it actually emits equal amounts of light all across the spectrum – far from it! – but only that it is a typical star to which other stars can be compared.

The **color index** of a star, B–V, is a quantitative measure of redness. It is zero for Vega, slightly negative for some of the hottest stars, and as high as 5 or 6 for some carbon stars. (B–I color indices are also used.) A B–V color index greater than 2 often indicates reddening by interstellar dust or carbon in the star.

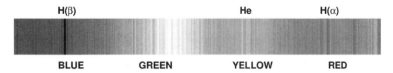

Figure 8.3. Spectrum of Altair (α Aquilae), showing absorption lines that reveal presence of hydrogen, helium, and other elements. (Maurice Gavin.)

The B_T and V_T magnitudes in the Hipparcos and Tycho catalogues were measured by the Tycho satellite's photometer, which does not exactly match the standard B and V passbands. B and V can be computed from B_T and V_T.

8.4.3 Spectroscopy

When the light of a star is spread out by a prism or diffraction grating, its smooth, continuous spectrum turns out to be interrupted by **absorption lines** (Figure. 8.3), places where light at a specific wavelength is weak or absent. The wavelengths indicate the chemical composition of the absorbing matter and hence of the star itself. Some stars also have **emission lines** indicating fluorescence in their outer layers.

Stars are predominantly made of hydrogen, but other elements show up clearly even when present only in small quantities. The hottest stars contain highly ionized atoms that can only exist at high temperatures; the coolest stars contain heavy elements and molecules.

Stellar spectra were originally classified alphabetically by strength of hydrogen lines, but around 1900 Antonia Maury of Harvard University rearranged them in order of temperature. With modern updates, the sequence is:

(hot, blue-white) **O – B – A – F – G – K – M – L – T** (cool, red)

This can be remembered by the undignified but unforgettable phrase "Oh be a fine girl/guy, kiss me *like this!*"

Stars near the O end are called **early-type** stars, and stars near the other end, **late-type**. This is a sequence of temperatures, not a sequence in time. Stars move along it in both directions as they age.

Each type is further subdivided by numbers: O0 ... O9, B0 ... B9, etc. (The Sun is type G2.) The suffix **e** means that emission lines are present, as in γ Cassiopeiae, type B0e.

Classes **C, R, N**, and **S** are alternatives to **M** with different chemical compositions. (**R** and **N** were consolidated to make **C**.) Classes **L** and **T** (recently introduced by J. D. Kirkpatrick and colleagues at Caltech) are very cool objects that can only be observed in infrared light; they are probably not stars at all, but rather "brown dwarfs" (p. 121).

At the other end of the range, **Wolf–Rayet** [pronounced *rah-YAY*] stars are more energetic than type O and are often designated **W**. They are thought to be stars that have lost their outer layers.

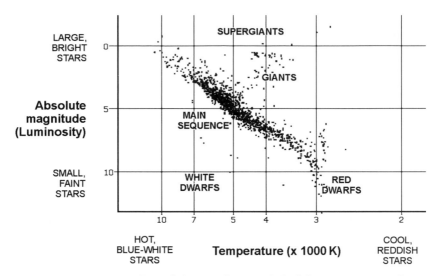

Figure 8.4. Hertzsprung–Russell diagram (scatter plot) of about 1200 stars in Cygnus, plotted with *Starry Night Pro* software. Red and white dwarfs are more common than the diagram suggests; they are too faint to see very far away.

8.5 Stellar physics

8.5.1 Mass, luminosity, and temperature

Most stars spend almost all of their lives as part of the **main sequence.** That means their light output (luminosity) and total size are related to their temperature in a simple way – bigger stars are hotter and brighter.

Figure 8.4 shows a plot of the temperatures and luminosities of about 1200 stars in the constellation Cygnus. The main sequence is the diagonal band along the middle of the diagram.

The main sequence is not a series of stages that stars go through; it is a set of types, and any particular star will remain at a specific point on it, with only minor shifting until, in its old age, it leaves the main sequence altogether.

Old age comes sooner for some stars than for others. Massive stars are hot, bright, and large and burn up their hydrogen fuel quickly. Low-mass stars never become so large or hot, and their supply of hydrogen lasts a long time.

8.5.2 Stellar evolution in brief

The **evolution** of a star is simply its aging process; stellar evolution has nothing to do with Darwinism or natural selection.

All stars begin life as thick spots in interstellar gas clouds made of nearly pure hydrogen. Once a thick spot gets started, its gravity attracts more matter, and it grows. Meanwhile, it heats up due to **gravitational contraction** – whenever you compress a gas, it gets hotter. The more massive the protostar is, the hotter it gets.

During the contraction phase of a star's development, it is large and relatively bright because of its large surface area, and its brightness varies irregularly. The star T Tauri is thought to be at this stage now.

What happens next depends on how much matter has accumulated and how hot it has managed to get. There are four possibilities:

1. If the star is very small, there will never be enough heat to start a fusion reaction. In that case, the object is a **brown dwarf** and not, strictly speaking, a star at all, though it glows from the heat acquired by contraction. Such objects are observable only with large telescopes using infrared light.
2. If the star is moderately small (e.g., one tenth the mass of the Sun), there will be enough heat to start a hydrogen fusion reaction, but the star will never get hot enough to fuse helium or heavier elements. In this case the star is a **red dwarf**, at the low-temperature, low-mass end of the main sequence. After a long time it will run out of fuel and cool down. Red dwarfs are by far the most common stars, but we can't see very many of them because of their low luminosity.
3. If the star is middle-sized (comparable to the Sun), it will fuse hydrogen for a long time and then switch to fusing helium. Its life cycle, from normal star to red giant and then white dwarf, is explained in more detail below.
4. If the star is very large (ten times the mass of the Sun), it will fuse hydrogen, then helium, and then heavier elements. Then, it will explode as a **supernova**, scattering heavy elements across the universe as it does so.

8.5.3 More about stellar evolution

The development of a middle-sized or large star, one that can switch from hydrogen to other fuels, is as follows.

Recall that the fusion reaction inside a star serves two purposes. Not only does it provide light and heat, it also provides pressure to keep the star from collapsing under its own gravity. Large stars require fusion energy to keep them large.

In a medium-sized or large main-sequence star, hydrogen fusion takes place mainly in the central core, and helium is its end product. Eventually, the core will fill up with helium, which is heavier than hydrogen, and the fusion reaction will begin to move into the outer layers. When this happens, the star will expand greatly, since its source of pressure is nearer the surface. Since a fixed amount of heat is spread over a larger area, the star cools and becomes a **red giant**. After expanding, it subsides somewhat.

The core of the red giant is still dense, of course, and at some point, the temperature and pressure will become high enough to start fusing helium. This stage is called the **helium flash** and is quite sudden (probably just a matter of seconds) but is not directly visible from outside the star. Its visible effect is that the star expands and becomes more luminous.

Eventually the red giant will run out of helium too, becoming full of carbon, which is the product of the helium reaction. A **carbon star** is one in which large amounts of this carbon are brought to the surface by convection, affecting the spectrum.

Either of two things happens next.

A middle-sized star is out of fuel at this point. As it is ending, the helium reaction moves out into the outermost layers and blows them into space; if a large amount of gas is dissipated in this way, we see it as a **planetary nebula**. What's left, in the middle, contracts tremendously because it no longer has an internal source of pressure. The result is a very hot, very dense **white dwarf**, heated only by gravitational contraction and therefore doomed to cool down and fade out.

A larger, hotter **supergiant** star, on the other hand, can start fusing heavier elements, especially carbon. This process doesn't last long, but it does provide lots of energy, and the star ends up exploding as a **supernova** and scattering its remnants across the universe.

Chapter 9
Double and multiple stars

9.1 The importance of double stars

More than half of all stars belong to double- or multiple-star systems. That is, they are gravitationally bound to other stars and orbit around the system's common center of gravity.

Amateur astronomers today have forgotten the excitement that accompanied the gradual discovery of this fact during the nineteenth century. Double stars provided the first opportunity to measure the mass of stars and to test the laws of physics outside the Solar System.

Many double stars show measurable orbital motion over just a few years. Many others need to be measured now, since determination of an orbit requires observations many years or centuries apart, and double-star work fell out of fashion in the twentieth century. The Hipparcos satellite made accurate measurements of numerous double stars in early 1991; even these observations are now old enough that it is worth while looking for subsequent changes.

It is often unknown whether a particular double is really a gravitationally bound **binary star** or merely an **optical double** comprising two unrelated stars in the same line of sight. In between are **common proper motion** pairs, stars that appear to be the same distance from Earth and have roughly the same proper motion, but in which orbital motion has not been observed.

A **visual binary** is a binary star whose orbit can be observed with telescopes. Less than a thousand orbits are known in any detail, and the available information about them is often inaccurate.

Other binary stars can be detected **spectroscopically** (because their spectra show Doppler shifts from orbital motion), **astrometrically** (by observing a wobble in the star's proper motion, indicating the gravitational pull of an unseen companion), or by determining that the star is an eclipsing binary (p. 134).

Double and multiple stars are strikingly attractive objects. Observing them requires steady air but not particularly dark skies; they can be viewed in moonlight and under urban conditions.

Figure 9.1. The double star Cor Caroli (α Canum Venaticorum, magnitudes 2.9 and 5.5, separation 19.4″, position angle 228°) is a fine sight in even the smallest telescopes. This CCD image resembles the telescopic view at medium power.

Many visual double stars show a striking color contrast. The obvious question is why two stars that have lived their whole lives together should be different colors – and the answer, in general, is that the larger star evolves faster and reaches the red giant stage sooner. That is why there are so many yellow-and-blue pairs like Albireo.

In other cases the color contrast is exaggerated by an optical illusion. When looking at two objects close together, the human eye tends to exaggerate any slight difference of color that may be present. Thus the companion of Antares looks greenish simply because Antares is so red.

9.2   Position angle and separation

A double star is described by giving the magnitudes of the two stars, the **separation** in arc-seconds (″), and the **position angle** (direction from the brighter star to the fainter one). Like azimuth, position angle is measured from north through east. Figure 9.2 shows how position angles look in telescopes with and without diagonals.

There can be more than two components, of course. Multiple stars include matched triples such as Σ 939, very mismatched triples such as Mizar, quadruples in a square arrangement such as θ^1 Orionis in M42, pairs of close pairs such as ϵ Lyrae, and irregular groups such as σ Orionis.

In a multiple-star system, the brightest component is designated A, and the companions are B, C, and so on, usually in order of increasing separation, but

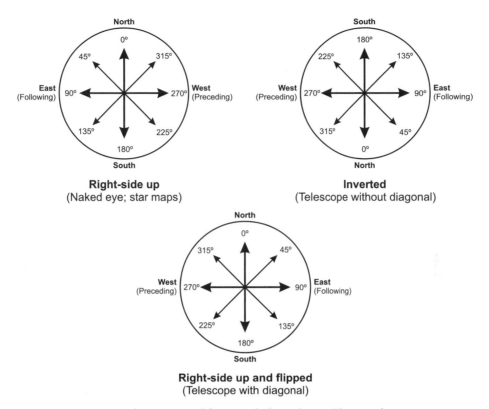

Figure 9.2. Position angle is measured from north through east. The top of picture is up when the telescope is aimed directly south. "Preceding" and "following" refer to diurnal motion, visible when motors are turned off.

sometimes in order of discovery. Figure 9.3 shows how position angles and separations are measured for a triple star.

9.3 Binary-star orbits

In real life, binary stars revolve around their common center of gravity, but the orbit as seen from Earth is always treated as an ellipse around the brighter star. As an example, Figure 9.4 shows the calculated orbit of Castor (α Geminorum) for the next few decades, based on data in the current *Washington Double Star Catalog* (WDS; see p. 131). Over 500 binary-star orbits are given in *Sky Catalogue 2000.0*, vol. 2, but those in WDS are often newer and more accurate.

The orbital elements are similar to Keplerian elements (p. 77). The period P is given in years; the inclination i is $0°$ for an orbit seen face-on and $90°$ for an orbit seen edge-on; the semi-major axis a is in arc-seconds; T is the date of periastron (as a fractional calendar year, such as 1963.25); and e, Ω, and ω describe the ellipse relative to the plane of the sky rather than the ecliptic. Periastron is the time when the stars are closest together and does not necessarily correspond to minimum angular separation seen from Earth.

Triple star Σ 939 in Monoceros

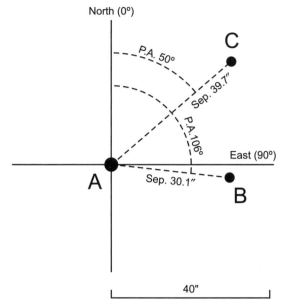

Figure 9.3. Position angle and separation are the fundamental measurements of a multiple-star system.

Castor (α Geminorum)
Orbital data from WDS

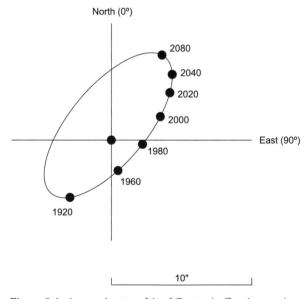

Figure 9.4. Approximate orbit of Castor (α Geminorum) as seen from Earth, calculated from elements in WDS (2001).

Table 9.1. *Dawes limit (for double stars of equal magnitude, barely split) and Rayleigh limit (for a somewhat cleaner split) as a function of telescope aperture*

Aperture	Dawes limit	Rayleigh limit (550 nm)
2.4 inches (60 mm)	1.9″	2.3″
3.5 inches (90 mm)	1.3″	1.6″
5 inches (12.5 cm)	0.9″	1.1″
6 inches (15 cm)	0.76″	0.93″
8 inches (20 cm)	0.58″	0.70″
10 inches (25 cm)	0.46″	0.56″

Figure 9.5. High-magnification view of a double star in perfectly steady air. Because of diffraction, star images are not points, but disks surrounded by rings.

To compute separation and position angle at any particular date, use the methods given in *Practical Astronomy With Your Calculator,* by Peter Duffett-Smith (Cambridge, 1988), but *do not assume that the elements are very accurate*; many of them are based on only a few observations.

9.4 Telescope limits

Double stars are a very stiff test of the quality of a telescope and the steadiness of the air. Some of the best observing conditions occur when the sky is hazy but the air is still. Double-star work is done at very high power; I have been known to use 700× on an 8-inch (20-cm) telescope.

The ability of a telescope to "split" double stars is limited by diffraction (Figure 9.5), which depends on telescope aperture. The diffraction limit has been quantified in two ways:

$$\textbf{Dawes limit} = \frac{4.56''}{\text{aperture (inches)}} = \frac{11.6''}{\text{aperture (cm)}}$$

$$\textbf{Rayleigh limit} \text{ (for 550 nm)} = \frac{5.5''}{\text{aperture (inches)}} = \frac{14''}{\text{aperture (cm)}}$$

The Dawes limit is based on the actual experience of double-star observer William Rutter Dawes (1799–1868) for pairs of equally matched stars about 6 magnitudes brighter than the limit of the telescope (e.g., 6th-magnitude stars in a 6-inch (15-cm) telescope). The Rayleigh limit is theoretical and depends on the wavelength of light. In practice, an observer needs some training to recognize stars as double when they are at the Dawes limit, even in perfect air, but stars at the Rayleigh limit are much easier to split.

When the components of a double star are not the same brightness – as is usually the case – these limits do not apply. A very rough formula that fits some observers' experiences is:

$$\text{Practical limit (arc-seconds)} = \frac{14 + 3.5(\Delta m)^2}{\text{aperture (cm)}}$$

where Δm is the difference in magnitudes. Thus, when the stars are 1.4 magnitudes apart, you can resolve stars only down to twice the Rayleigh limit.

9.5 Making measurements for yourself

9.5.1 The need for measurements

Double-star work is one area of astronomy where the human eye has not been replaced by photographic detectors. The reason is that an experienced human observer can take advantage of moments of unusually steady air and can make judgements about images at the limit of resolution.

Many double stars have not been measured accurately in decades, and new measurements today would reveal important information about their orbits. The *Washington Double Star Catalog* web page (http://ad.usno.navy.mil/ad/wds) includes lists of doubles that need to be measured, with finder charts, and double-star work is coordinated by the B.A.A. (p. 30). Even a 4-inch (10-cm) telescope is big enough for useful work. The ability to measure position angle and separation is also useful for precise location of other celestial objects.

9.5.2 Teague's reticle method

Traditionally, in order to measure position angle, you must have an equatorially mounted telescope to establish which way is north. But in *Sky & Telescope*, July, 2000, Thomas Teague outlined a method that works equally well in altazimuth mode. The key idea is that all stars drift directly west when the drive motor is turned off; thus the Earth rather than the telescope determines the compass points.

Teague's method relies on a micrometric eyepiece with a reticle like that shown in Figure 9.6. Eyepieces of this type are made by both Meade and Celestron; the Celestron version has more eye relief and is the one I recommend.

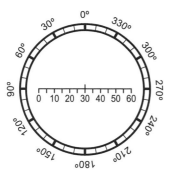

Figure 9.6. Micrometric reticle for measuring double stars. Celestron and Meade make eyepieces of this type, with slight variations.

9.5.3 Calibrating the linear scale

In the center of the reticle is a linear scale from 0 to 60. These units are arbitrary and mean nothing until you calibrate the eyepiece as follows. Turn the eyepiece until the scale runs exactly east–west, place a star on the eastern end, turn off the drive, and time how long it takes the star to reach the other end. (Usually, this takes several tries, since you won't have the scale lined up in exactly the right direction at first.) Then use the formula:

$$\text{Length of scale (arc-seconds)} = \text{Drift time (seconds)} \times 15.041 \cos \delta$$

where δ is the declination of the star. Here $15.041''$ is the amount of diurnal motion in one second of time (it would be exactly $15''$ in one *sidereal* second).

Teague recommends using a star with high declination, around $60°$ or $70°$, so that the drift will be slow, and performing the calibration repeatedly, then averaging the results. Once you know the length of the scale, you can interpret the units into which it is divided; obviously 23 on a scale of 60 is 23/60 of the total length.

This calibration applies only as long as you are using exactly the same telescope and diagonal. Any change in the optical system – even removing your eyeglasses and refocusing – will throw it off. You will probably want to do two calibrations, one with a Barlow lens and one without. Along the way, you may discover that your $2\times$ Barlow is actually $1.85\times$, or $2.43\times$, or whatever.

9.5.4 Taking a measurement

Figure 9.7 shows how a measurement is actually taken. First rotate the eyepiece to line up the double star on the linear scale, with the brighter star at $0°$, and read off the separation. Then, without rotating the eyepiece further, slew the telescope so that a star is exactly in the center, and turn off the drive. Watch where the star crosses the position-angle scale, and read off the angle. Always *write down the numbers you actually see*, and interpret them later; don't try to do arithmetic while observing.

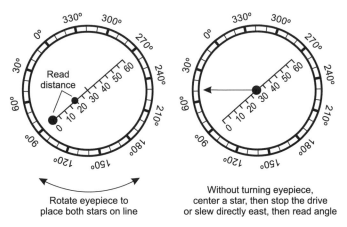

Figure 9.7. How to take a measurement using Teague's method. Always write down the numbers you actually see, and interpret them afterwards.

The previous section explained how to interpret the separation scale. The position-angle scale also requires some interpretation. The scale reading will be off by 90° because the scale starts at the west instead of north. Because several kinds of reticles are in use, I can't give instructions to be followed blindly, but two formulae you can try are:

Position angle = 90° + reading   (if the scale reads the right way)

Position angle = 90° + (360° − reading)   (if the scale reads backwards)

In either case, add 360° if the result is negative, and subtract 360° if the result exceeds 360°. Measuring one or two known doubles will quickly tell you which formula applies to your equipment.

9.5.5 Turning off the drive motors

Most computerized telescopes provide a way to turn the drive motors off and on again without losing alignment. In equatorial mode, simply slew eastward at 1× sidereal rate (the slowest slewing speed). In that special situation, the motors simply turn off, and the drift that you see is caused only by the Earth, not the motors, so its direction is not affected by the accuracy of your polar alignment.

Other techniques to try include switching to land mode on any Meade telescope; turning tracking off on the NexStar; or using the "Sleep Scope" utility on the Autostar. Whether you will have to realign afterward depends on the telescope model and firmware version.

9.6 Multiple-star nomenclature and catalogues

Double-star observing started with a set of nineteenth-century observers, each of whom published a list of a few dozen or a few hundred stars. Table 9.2 shows the

Table 9.2. *Abbreviations for the names of some early double-star observers. The first abbreviation is used in print; the second, in computer files*

β	BU	S. W. Burnham (1838–1921)
Δ	DUN	J. Dunlop
Σ	STF	F. G. W. Struve (1793–1864) [pronounced *SHTROO-va*]
Es	ES	T. E. Espin (1858–1934)
H	H	Sir William Herschel (1739–1822)
h	HJ	Sir John Herschel (1792–1871)
Kr	KR	A. Krüger (Krueger)
OΣ, OΣΣ	STT	Otto Struve (1819–1905) (2 volumes)
Wnc	WNC	F. A. Winnecke [pronounced *VIN-eck-a*]

abbreviations for a few of the old-time observers' names; many more are listed in *Sky Catalogue 2000.0*. Each abbreviation is followed by a catalogue number to designate a particular star. For instance, Castor (α Geminorum) is also Σ 1110.

The current authority on double and multiple stars is the *Washington Double Star Catalog* (WDS), maintained online by the U.S. Naval Observatory at http://ad.usno.navy.mil/ad/wds and updated continuously.[1] WDS catalogue numbers are of the form 07346+3153, which directly encodes the position of the star, R.A. $07^h34.6^m$, declination $+31°53'$ (epoch 2000.0). Catalogue entries include positions, magnitudes, orbits (with the ability to plot an orbit against a large data set of observations), and cross-indexing information.

Before WDS, the definitive catalogue was IDS (*Index Catalogue of Visual Double Stars*, by H. M. Jeffers and W. H. van den Bos, Lick Observatory, 1963), and before that, ADS ("Aitken's double stars"), i.e., the *New General Catalogue of Double Stars*, by R. G. Aitken (Carnegie Institution, 1932).[2] Aitken's numbers are often marked with A, such as A6175 = ADS 6175 (Castor).

Our knowledge of multiple stars grows by accumulation, and the latest catalogues combine all the reliable information available in earlier ones. Clearly, the oldest observations are often the most valuable for determining orbits. Be forewarned, however, that many of the separations and position angles in Smyth's *A Cycle of Celestial Objects* (1844) are approximate; do not use them as a basis for long-term comparison.

---

[1] I was fortunate enough to be able to provide a couple of updates to WDS while preparing the object list in Part IV of this book.

[2] Not to be confused with Dreyer's *New General Catalogue* (NGC) of clusters, nebulae, and galaxies (p. 158).

Chapter 10
Variable stars

10.1 Overview

All stars are variable – it's just that some of them have not varied appreciably during human history. Every star changes brightness as it ages, and many stars *pulsate* – that is, they get brighter and dimmer in a regular cycle.

Besides being interesting to watch, variable stars provide opportunities for amateurs to contribute to scientific knowledge. Much of the year-by-year monitoring of variable stars is done by amateurs, and amateurs discover many novae and supernovae. A large telescope is not required; some of the most productive observers use small, wide-field instruments or even binoculars.

Amateur variable-star work is coordinated by the B.A.A. (p. 30) and the American Association of Variable Star Observers (AAVSO, 25 Birch St., Cambridge, MA 02138, U.S.A., http://www.aavso.org). The AAVSO maintains a web site with current information on thousands of stars and issues bulletins about unexpected phenomena.

Besides AAVSO and B.A.A. training materials, two very good guides to variable-star observing are the book *Observing Variable Stars,* by David H. Levy (Cambridge University Press, 1998), and the chapter on variable stars by M. Dumont and J. Gunther in Patrick Martinez' *The Observer's Guide to Astronomy,* vol. 2, pp. 775–846. Levy's book is designed for beginners and casual skygazers; Dumont and Gunther's treatment of the subject is more technical.

10.2 Types of variables

10.2.1 Pulsating variables

A **pulsating** variable star is one that alternately expands and contracts. These are stars that never achieve a stable balance between gravity, which holds the star together, and heat, which keeps it pressurized and prevents collapse.

These stars are unstable because they contain matter whose opacity varies with pressure. When the star expands, the atoms change their ionization state and the gas becomes more transparent; then they can no longer trap as much heat, so the star cools down and contracts again. When the star contracts, the opposite change takes place; the same gas becomes more opaque, absorbs more heat, and gets ready to expand again. The change of ionization takes time; this introduces a delay and causes the star to go into a repeating cycle instead of reaching equilibrium.

There are many kinds of pulsating variables, and no two star catalogues use quite the same classification. In general, the larger the **amplitude** (amount of variation), the larger the star and the longer the period.

Cepheids [pronounced *SEE-fee-ids*] are white supergiants with periods of a few days, amplitudes of one or two magnitudes, and relatively good predictability. The prototype for this class is δ Cephei (p. 224).

The luminosity (absolute magnitude) of a Cepheid correlates very closely with the period. This fact was discovered in 1912 by Henrietta Leavitt, who studied Cepheids in the Large Magellanic Cloud, all of which are approximately the same distance from us. The "Cepheid yardstick" provides a way to determine the distance of any galaxy in which individual Cepheids can be observed. There are two kinds of Cepheids, **classical Cepheids** and **W Virginis stars**, with slightly different period–luminosity relations.

RR Lyrae stars have shorter periods (as little as an hour) and amplitudes less than one magnitude, sometimes much less. The prototype RR Lyrae star is of course RR Lyrae (magnitude 7.1–8.1, period 13 hours 40 minutes). Another interesting RR Lyrae star is CY Aquarii (p. 231), which combines a relatively large amplitude with a very short period; it varies from magnitude 10.4 to 11.2 with a period of only 88 minutes.

Dwarf Cepheids (δ **Scuti stars**) are another class of low-amplitude, short-period pulsating variables. They often show simultaneous pulsations at more than one frequency.

Mira-type (long-period) variables are red giants with periods of 80 to 1000 days and amplitudes greater than 3 magnitudes. The most famous star of this type is Mira (*o* Ceti), which varies between magnitudes 3 and 9 in a slightly irregular cycle that averages 332 days (Figure 10.1; see also p. 242). Long-period variables are redder at minimum than at maximum.

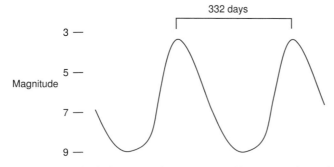

Figure 10.1. The light curve of Mira, a typical long-period variable.

A **semiregular** variable is a long-period variable whose cycle is somewhat hesitant and halting. An example is V Aquilae (p. 216), magnitude 6.6–8.4, average period 353 days. Red giant variables tend to be classified as semiregular even if little regularity is evident, as in the case of Betelgeuse (α Orionis), which fluctuates between magnitudes 0.4 and 1.3.

10.2.2 Irregular variables

Irregular variables (**nebular variables**) are immature, newly formed stars that have just condensed out of gas clouds. An example is T Tauri (mag. 8.4–13.5), which is normally close to magnitude 10 and sometimes brightens or fades over a period of weeks.

At other times T Tauri flickers rapidly from minute to minute over a range of perhaps 0.5 magnitude. The AAVSO recommends making several estimates of T Tauri each evening, with the same comparison stars, so that if one of these odd episodes is going on, you will catch it. T Tauri is embedded in Hind's Variable Nebula, NGC 1555.

R Coronae Borealis variables are stars that stay bright most of the time but occasionally fade. The prototype is of course R Coronae Borealis. Normally it is 6th magnitude, but it can drop to 14th magnitude over just a few days, then recover slowly over a period of months. This is thought to be due to a buildup of carbon in the star's outer layers. Another star of the same type is RY Sagittarii (magnitude 6.5–14).

A **flare star** is a star that has occasional flares just like the Sun, except that the star is so small that the flare adds appreciably to its brightness. One component of the double star Krüger 60 (p. 224) is an example.

10.2.3 Eclipsing binaries

Some of the most famous variable stars aren't really variable at all. They are stars whose light is periodically blocked by a fainter companion star orbiting around the main star. These are called **eclipsing binaries** and the most famous example is Algol (β Persei), which dims from magnitude 2.1 to magnitude 3.4 for a couple of hours every 2.8673 days with clockwork precision (Figure 10.2).

Predicted times of minima of Algol are published in *Sky & Telescope*, in the *Observer's Handbook* of the Royal Astronomical Society of Canada, and in the *Handbook of the British Astronomical Association*. The star's name, Arabic for "the ghoul," suggests that even in ancient times, star gazers knew there was something spooky about it.

Another famous eclipsing binary is BM Orionis (θ^1 Orionis B), the faintest of the four Trapezium stars within the Orion Nebula (Figure 18.3, p. 243). It drops from magnitude 7.9 to 8.5 for nine hours every 6.47 days. Predictions are

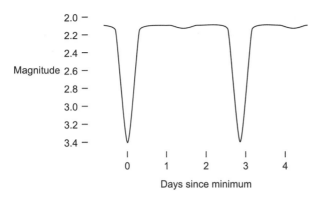

Figure 10.2. The light curve of an eclipsing binary (Algol).

sometimes published in *Sky & Telescope*. The brightest Trapezium star, θ^1 Orionis A, is also an eclipsing binary; it dims from magnitude 6.7 to 7.7 once every 65.4 days.

Because they are so regular, there is little to be learned from observing eclipsing binaries that are already well known. But more eclipsing binaries are waiting to be discovered. Each eclipse is short, but there is no limit to how long the period between eclipses can be. Thus, many eclipsing binaries have never been observed in eclipse, or have been observed so few times that the period is not known. The variability of the brightest Trapezium star was not discovered until 1975, even though amateur astronomers had been staring at it regularly for two centuries.

We only see these eclipses because we are looking at the binary system from the correct angle, of course. Any number of other stars in the sky are equally close binaries that do not eclipse each other as seen from Earth.

10.2.4 Novae

Another type of very close double can become a **nova**. What happens is that a white dwarf orbits so close to a normal star that it can refuel itself, pulling matter from the main star. Periodically, hydrogen fusion starts up and the white dwarf is suddenly a normal star again – or at least its outermost layers are. The sudden increase in brightness lasts only a few days and the nova reverts to its former state.

Note that novae are unrelated to supernovae (p. 121), and that an ordinary star, by itself, cannot "go nova" – only a close-orbiting white dwarf can do that. It is thought that all novae are recurrent, although the interval between recurrences may be many thousands of years. Some of them, such as SS Cygni, recur every few months and are called **dwarf novae** because of their short period and relatively small amplitude (just a few magnitudes).

What novae and supernovae have in common is that they are both character-ized by a very rapid rise of 10 magnitudes or more, which is rarely observed, followed by a decline that takes days, weeks, or months.

Bright novae are rare, but seventh- or eighth-magnitude novae in rich fields of the Milky Way occur every year; the hard part is finding out about one while it's still around to be observed. For rapid notification, see http://www.aavso.org or subscribe to the AAVSO's e-mailed bulletins.

There has not been a supernova in our galaxy since 1604, but one could occur at any time. Supernovae in thousands of *other* galaxies are within reach of amateur telescopes; they typically reach twelfth to fifteenth magnitude as seen from Earth and can be photographed with CCD cameras. Some of the brighter ones are noticeable visually.

10.2.5 Reporting a discovery

If you discover a nova or supernova, you're caught in a difficult situation – the discovery should be reported immediately so that the rise, if it is still in progress, can be observed, but first you have to make sure your discovery is genuine. Plenty of other things can look like a nova – bright planets, geostation-ary satellites, asteroids, and stars that just happen to be missing from a chart that should show them. The last of these are amazingly numerous.

The first thing to do is make absolutely sure of the position of "your" star, and, if possible, photograph it. If it's a nova still on the rise, make estimates and/or take pictures every few minutes. *Record the exact time of every observation.* As you do this, make sure the star isn't moving. Asteroids and satellites move relative to the surrounding stars (though an asteroid can stand still at certain points in its orbit).

You can rule out all known asteroids by using the interactive web site at http://cfa-www.harvard.edu/iau/cbat.html (the International Astronomical Union Central Bureau for Astronomical Telegrams). Give the computer the exact position of the nova or the NGC number of the galaxy, and it will tell you what asteroids are known to be there, if any.

In theory, you can report your discovery directly to the IAU at the web site just mentioned, but in practice, that is not recommended. Instead, get your observation confirmed by someone else, preferably another amateur astronomer in your own neighborhood, or someone reached through the B.A.A. or AAVSO. Then, make the report with expert help. The advice on p. 71 about reporting discoveries of comets is applicable.

The IAU has received so many mistaken reports of supernovae in distant galaxies that observers are now expected to do three of the following four things before reporting one:

1. measure its position precisely (relative to known stars on a photograph or CCD image, not just the readout of the telescope);

2. observe it on a second night to verify that it hasn't moved;
3. show that the object was not present on photographs taken earlier (not necessarily your own photographs, but reliable images with an equal or deeper magnitude limit); and/or
4. confirm spectroscopically that the object is a nova or supernova.

Usually, item 4 is beyond the reach of amateurs, but the other three are feasible, particularly if you can get help from others. Naturally, a really bright nova requires less checking than a faint one.

Discoveries of ordinary variable stars are not quite so urgent. The best tactic is to make repeated observations over several days or more, meanwhile checking all available charts and star catalogues, and finally communicating with the B.A.A. or AAVSO. Plenty of variable stars are waiting to be discovered, but there are also plenty of weeds amid the crops.

10.3 Nomenclature

10.3.1 Letter designations

The variable stars in each constellation are designated R, S, T... in order of discovery, unless they already have well-known Greek-letter designations. This system was started by F. W. A. Argelander in the mid-1800s; it begins at R because Bayer and Lacaille letters never go past Q. After Z, the designations are RR...RZ, SS...SZ, on through ZZ and then AA...QZ (skipping J). After using up all 334 combinations, the sequence continues with V335, V336, etc. By early 2001, Sagittarius was up to V4643 and still counting.

10.3.2 Harvard designations

The AAVSO identifies variable stars by a system of six-digit numbers that started at Harvard University, giving the right ascension in hours and minutes and the declination in degrees. For example, $0942+11$ (for $09^h42^m +11°$) denotes R Leonis. The plus or minus is sometimes omitted; when this is done and the declination is negative, the last two digits are printed in italics, boldface, or underlined. When two variable stars have the same Harvard designation, letters A, B, C (etc.) are added at the end of the designation to distinguish them.

The catch is that the coordinates are epoch 1900, a century behind today's star catalogues. To find a star's Harvard designation, you can use the table of precession from 1950 to 2000 on p. 249, but apply the corrections *backward* (subtracting instead of adding) and apply them *twice* (for 100 years instead of 50). That will take the coordinates from 2000 to 1900. The Harvard rules for rounding to the nearest whole number are somewhat complex and are explained in the AAVSO's literature.

10.3.3 GCVS numbers

The "master catalogue" of variable stars is the *General Catalogue of Variable Stars* (*GCVS*), edited by P. N. Kholopov (earlier editions by B. V. Kukarkin *et al.*), Moscow, 1985, now continuously updated online at http://www.sai.msu.su.

The GCVS uses the letter designations begun by Argelander, but GCVS designations also have a numerical form, which is used by Meade LX200 telescopes. Each GCVS number consists of two digits for the constellation and four digits for the star, as shown in the following charts. Thus R Lyrae is 520001 (constellation 52, variable star 1).

GCVS constellation codes

The full GCVS number consists of the constellation code followed by the star designation (see below).

01 Andromeda	31 Cygnus	61 Pavo
02 Antlia	32 Delphinus	62 Pegasus
03 Apus	33 Dorado	63 Perseus
04 Aquarius	34 Draco	64 Phoenix
05 Aquila	35 Equuleus	65 Pictor
06 Ara	36 Eridanus	66 Pisces
07 Aries	37 Fornax	67 Piscis Austrinus
08 Auriga	38 Gemini	68 Puppis
09 Boötes	39 Grus	69 Pyxis
10 Caelum	40 Hercules	70 Reticulum
11 Camelopardalis	41 Horologium	71 Sagitta
12 Cancer	42 Hydra	72 Sagittarius
13 Canes Venatici	43 Hydrus	73 Scorpius
14 Canis Major	44 Indus	74 Sculptor
15 Canis Minor	45 Lacerta	75 Scutum
16 Capricornus	46 Leo	76 Serpens
17 Carina	47 Leo Minor	77 Sextans
18 Cassiopeia	48 Lepus	78 Taurus
19 Centaurus	49 Libra	79 Telescopium
20 Cepheus	50 Lupus	80 Triangulum
21 Cetus	51 Lynx	81 Triangulum Australe
22 Chamaeleon	52 Lyra	82 Tucana
23 Circinus	53 Mensa	83 Ursa Major
24 Columba	54 Microscopium	84 Ursa Minor
25 Coma Berenices	55 Monoceros	85 Vela
26 Corona Australis	56 Musca	86 Virgo
27 Corona Borealis	57 Norma	87 Volans
28 Corvus	58 Octans	88 Vulpecula
29 Crater	59 Ophiuchus	
30 Crux	60 Orion	

GCVS star numbers

The full GCVS number consists of the constellation code (above) followed by the star number.

0001 R	0041 VW	0081 BC	0121 CU	0161 ER	0201 GS	0241 IX
0002 S	0042 VX	0082 BD	0122 CV	0162 ES	0202 GT	0242 IY
0003 T	0043 VY	0083 BE	0123 CW	0163 ET	0203 GU	0243 IZ
0004 U	0044 VZ	0084 BF	0124 CX	0164 EU	0204 GV	0244 KK
0005 V	0045 WW	0085 BG	0125 CY	0165 EV	0205 GW	0245 KL
0006 W	0046 WX	0086 BH	0126 CZ	0166 EW	0206 GX	0246 KM
0007 X	0047 WY	0087 BI	0127 DD	0167 EX	0207 GY	0247 KN
0008 Y	0048 WZ	0088 BK	0128 DE	0168 EY	0208 GZ	0248 KO
0009 Z	0049 XX	0089 BL	0129 DF	0169 EZ	0209 HH	0249 KP
0010 RR	0050 XY	0090 BM	0130 DG	0170 FF	0210 HI	0250 KQ
0011 RS	0051 XZ	0091 BN	0131 DH	0171 FG	0211 HK	0251 KR
0012 RT	0052 YY	0092 BO	0132 DI	0172 FH	0212 HL	0252 KS
0013 RU	0053 YZ	0093 BP	0133 DK	0173 FI	0213 HM	0253 KT
0014 RV	0054 ZZ	0094 BQ	0134 DL	0174 FK	0214 HN	0254 KU
0015 RW	0055 AA	0095 BR	0135 DM	0175 FL	0215 HO	0255 KV
0016 RX	0056 AB	0096 BS	0136 DN	0176 FM	0216 HP	0256 KW
0017 RY	0057 AC	0097 BT	0137 DO	0177 FN	0217 HQ	0257 KX
0018 RZ	0058 AD	0098 BU	0138 DP	0178 FO	0218 HR	0258 KY
0019 SS	0059 AE	0099 BV	0139 DQ	0179 FP	0219 HS	0259 KZ
0020 ST	0060 AF	0100 BW	0140 DR	0180 FQ	0220 HT	0260 LL
0021 SU	0061 AG	0101 BX	0141 DS	0181 FR	0221 HU	0261 LM
0022 SV	0062 AH	0102 BY	0142 DT	0182 FS	0222 HV	0262 LN
0023 SW	0063 AI	0103 BZ	0143 DU	0183 FT	0223 HW	0263 LO
0024 SX	0064 AK	0104 CC	0144 DV	0184 FU	0224 HX	0264 LP
0025 SY	0065 AL	0105 CD	0145 DW	0185 FV	0225 HY	0265 LQ
0026 SZ	0066 AM	0106 CE	0146 DX	0186 FW	0226 HZ	0266 LR
0027 TT	0067 AN	0107 CF	0147 DY	0187 FX	0227 II	0267 LS
0028 TU	0068 AO	0108 CG	0148 DZ	0188 FY	0228 IK	0268 LT
0029 TV	0069 AP	0109 CH	0149 EE	0189 FZ	0229 IL	0269 LU
0030 TW	0070 AQ	0110 CI	0150 EF	0190 GG	0230 IM	0270 LV
0031 TX	0071 AR	0111 CK	0151 EG	0191 GH	0231 IN	0271 LW
0032 TY	0072 AS	0112 CL	0152 EH	0192 GI	0232 IO	0272 LX
0033 TZ	0073 AT	0113 CM	0153 EI	0193 GK	0233 IP	0273 LY
0034 UU	0074 AU	0114 CN	0154 EK	0194 GL	0234 IQ	0274 LZ
0035 UV	0075 AV	0115 CO	0155 EL	0195 GM	0235 IR	0275 MM
0036 UW	0076 AW	0116 CP	0156 EM	0196 GN	0236 IS	0276 MN
0037 UX	0077 AX	0117 CQ	0157 EN	0197 GO	0237 IT	0277 MO
0038 UY	0078 AY	0118 CR	0158 EO	0198 GP	0238 IU	0278 MP
0039 UZ	0079 AZ	0119 CS	0159 EP	0199 GQ	0239 IV	0279 MQ
0040 VV	0080 BB	0120 CT	0160 EQ	0200 GR	0240 IW	0280 MR

0281 MS	0290 NO	0299 NX	0308 OU	0317 PS	0326 QR	0335 V335
0282 MT	0291 NP	0300 NY	0309 OV	0318 PT	0327 QS	0336 V336
0283 MU	0292 NQ	0301 NZ	0310 OW	0319 PU	0328 QT	etc.
0284 MV	0293 NR	0302 OO	0311 OX	0320 PV	0329 QU	
0285 MW	0294 NS	0303 OP	0312 OY	0321 PW	0330 QV	
0286 MX	0295 NT	0304 OQ	0313 OZ	0322 PX	0331 QW	
0287 MY	0296 NU	0305 OR	0314 PP	0323 PY	0332 QX	
0288 MZ	0297 NV	0306 OS	0315 PQ	0324 PZ	0333 QY	
0289 NN	0298 NW	0307 OT	0316 PR	0325 QQ	0334 QZ	

10.4 Observing techniques

10.4.1 Estimating magnitudes

Figure 10.3 shows one of the AAVSO's charts for observers of SS Cygni (and, as it happens, also V1668 Cygni, which is nearby). The numbers indicate accurately known star magnitudes with the decimal point left out. Thus "118" denotes $m = 11.8$. Underlining indicates that some of the magnitudes are from a particular set of photometric observations indicated at the bottom of the chart.

The most obvious way to estimate a magnitude is by **direct interpolation**. If the brightness of the variable star is halfway between comparison stars of magnitudes 11.0 and 11.8, it must be 11.4. With a good set of comparison stars, estimating to a precision of 0.2 magnitude is easy, and 0.1-magnitude precision can be obtained with practice.

Particularly in Europe, the **step method** is popular. Instead of magnitudes, the comparison stars are labeled with arbitrary designations (A, B, C, etc.). This helps keep the observer honest, and it also makes it possible to observe without a chart of accurate magnitudes – just label the stars any way you find convenient, and look them up afterward. In that situation, you should use several sets of comparison stars because some of them may turn out to be variable.

One "step" is the least perceptible difference in brightness; two steps are a difference that is perceptible but not prominent; three steps are a noticeable difference, and so forth.

The technique is to estimate the difference, in steps, between the variable star and at least two comparison stars, one fainter than the variable and one brighter. Call these stars A and B respectively. Then

A(2)V

means the variable is two steps brighter than A, and

V(3)B

means the variable is three steps fainter than B. Put them together, and you get

A(2)V(3)B

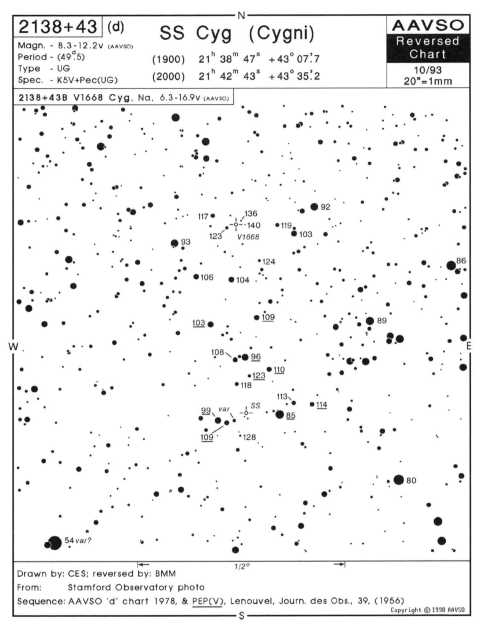

Figure 10.3. SS Cygni. AAVSO "D" scale chart, reversed to match the view through a telescope with a diagonal.

indicating that the star is two-thirds of the way along the magnitude scale from A to B.

It goes without saying that the magnitudes of the comparison stars must be accurate. *Do not use magnitudes from the Hubble Guide Star Catalog;* they are erratic. Hipparcos V (visual) magnitudes are better, but AAVSO charts should be used if at all possible. As in Figure 10.3, footnotes on the charts indicate who did the photometric measurements.

10.4.2 Telescope considerations

Computerized telescopes are very helpful in variable star work because they get you to the desired star quickly. However, it is still necessary to use star charts for positive identification and to identify comparison stars. Bear in mind that in a telescope with a diagonal, the image is flipped left-to-right relative to ordinary AAVSO telescopic charts (those with south up). "Reversed" charts, for observers with diagonals, are available.

A telescope is most useful with variable stars about two or three magnitudes above its magnitude limit as given in Table 8.1 (p. 114). If the star is too bright, not only is its magnitude hard to judge, but there won't be a comparison star within the same field. Stars brighter than 6th magnitude require binoculars; those brighter than about 4th magnitude can only be observed with the naked eye. The brightest variables, such as Betelgeuse, are hard to observe because there is often no suitable comparison star anywhere in the sky!

Most telescopes suffer a certain amount of **vignetting** – that is, the image is fainter at the edges than in the center of the field. For most purposes, this is not a problem, since the vignetting is too slight to be visible. In fact, some vignetting is inevitable if the telescope is well shielded against internal reflections.

However, even slight vignetting can throw off subtle magnitude estimates. The obvious remedy is to place both stars equal distances from the center of the field when comparing them.

One curable source of vignetting is a diagonal prism or mirror whose clear aperture is too small. There are "wide-field" diagonals that use almost the full diameter of the tube. To completely absolve the diagonal from suspicion, you can use a 2-inch diagonal with $1\frac{1}{4}$-inch eyepieces.

Also, look out for eyepieces that do not bring peripheral stars into sharp focus. It is often easier to throw the entire field slightly out of focus than to try to get all of it perfectly sharp; you get valid comparisons either way.

10.4.3 Sources of difficulty

Never let your preconceptions influence your magnitude estimates. It is better not to remember any predictions or even your own previous estimates of the same star. The best variable-star observers do dozens of estimates every night, so that they can't remember their own data from the previous night. Do not worry about being consistent; it is better to have random errors than to replicate your own previous mistakes.

Short-period fluctuations in brightness do occur, particularly in T Tauri type variables. If a star seems brighter or fainter than ten minutes ago, record another estimate. Both estimates may be right.

Long-period variables are usually reddish, sometimes extremely so. Comparing a red star to a white one can be tricky. Prolonged staring at a red star makes it seem to grow brighter, so comparisons have to be made quickly.

What's worse, the sensitivity of the eye to red light depends on the overall light level. In complete darkness, the eye is relatively insensitive to red. Under moonlight or city lights, or when you have just gone outdoors and your eyes are not dark-adapted, red stars will seem *brighter* (relative to their white companions) than under better viewing conditions.

Finally, to avoid errors, always record the *local* date and time of your observations, not just the Julian date, and always record local time, not just UT. Calculations done in the field are error-prone.

10.4.4 Photographic observation

You can observe variable stars by photographing them and then, later, comparing the variable to other stars on the same photograph. Black-and-white panchromatic films such as Tri-X Pan and T-Max 100 have roughly the right spectral response for V (visual) magnitudes. Color films are much more sensitive to red than the human eye. Some CCD cameras can output photometric magnitudes directly, but a filter is likely to be necessary to cut down the camera's excessive response to red and infrared light. Of course, if what you're doing is plotting a light curve, spectral response is less important as long as all the observations are made with the same instrument.

Chapter 11
Clusters, nebulae, and galaxies

11.1 The lure of the deep sky

The "deep sky" is the sky beyond the Solar System, and while it theoretically includes all stars except the Sun, in practice "deep-sky observing" means observing star clusters, nebulae, and galaxies.

For many amateur astronomers, including myself, deep-sky observing is the most interesting specialty, the most far-ranging form of celestial sightseeing. We bought our telescopes in order to see the universe. For the variety of sights and the variety of astrophysical processes behind them, deep-sky observing is unsurpassed.

Critics point out that deep-sky observers are unlikely to contribute anything to science, since most of the objects are near the limit of visibility in amateur telescopes, and apart from occasional supernovae in distant galaxies, there is nothing that amateur equipment can discover. That doesn't deter us. Seeing the sights is enough.

11.2 Deep-sky objects

11.2.1 Asterisms

An asterism is any small group of stars that catches the eye, whether or not the stars form a cluster in space. M73, for instance, is an asterism of four stars in Aquarius.

In recent years, amateur astronomers have given colorful names to numerous telescopic asterisms, many of which were enumerated by Philip S. Harrington in *The Deep Sky: An Introduction* (Sky Publishing, 1998). That book is actually an excellent overall guide to deep-sky observing and includes a small but useful star atlas. I must confess to having subsequently named two asterisms myself, Webb's Horseshoe (p. 207) and the Perfect Right Angle (p. 245).

11.2.2 Open clusters

An **open cluster** is a loose grouping of stars within our galaxy. Striking examples include the Pleiades, the Double Cluster in Perseus, and M7. The stars of a cluster are genuinely close together in space, though not extremely close; if you lived in one, your night sky would contain plenty of first-magnitude stars, but the rest of the universe would still be visible.

The stars in an open cluster are not gravitationally bound and will eventually separate. Sometimes a cluster can be recognized only by the common proper motion of its components; it is then called a **moving cluster**. The stars $\beta, \delta, \gamma, \epsilon$, and ζ Ursae Majoris – that is, most of the Big Dipper – constitute a moving cluster.

Open clusters are sometimes called **galactic clusters** because most of them are close to the plane of our galaxy (the Milky Way). (Galactic clusters are not the same as clusters *of* galaxies, of course.) One open cluster is a constellation, Coma Berenices.

Globular clusters are huge, dense, round concentrations of stars, typically 50 light-years in diameter and so densely packed that if you lived in one, you would have a hard time seeing anything outside it. Globular clusters form a loose halo around our galaxy, mainly outside its principal plane. If you envision our galaxy as a pancake, the globular clusters are like a swarm of flies hovering above and below it. Prominent globulars include ω Centauri, M22, and M13.

11.2.3 Nebulae

Nebulae are clouds of interstellar gas and/or dust. Real nebulae are much thinner than those on *Star Trek;* many are as thin as the best laboratory vacuums on Earth.

Reflection nebulae shine by reflecting starlight; they photograph as white or blue. **Emission nebulae** are flourescent; they glow the way a neon sign glows, except that the hydrogen gas is energized not by electricity, but by radiation from nearby stars. Emission nebulae look grayish to the eye but photograph as brilliant red on most color films. The Trifid Nebula (M20) has regions of both kinds and shows up on photographs as a pattern of red, white, and blue. **Dark nebulae** do not glow at all because there are no stars close enough to illuminate them; they are seen as shadows blocking the light of more distant stars.

Classifying nebulae another way, **diffuse nebulae** are large clouds of interstellar gas; **planetary nebulae** (called "planetary" because they are round) are the compact, bright emission nebulae thrown off by expiring stars (p. 122); and **supernova remnants** are the larger nebulae thrown off by exploding supernovae. Our sky presently includes one fresh supernova remnant (the Crab Nebula, M1, which dates from 1054 A.D.) and many loose fragments of older ones.

11.2.4 Our galactic neighborhood

Galaxies are the only celestial objects that always look better in long-exposure photographs than in the telescope. The reason is that, though immense in space, galaxies are invariably faint, except for the central region. In fact, the most common astronomical error in science fiction films is to show a bright galaxy looming up, like a Palomar photograph, outside the spaceship. In reality, like all other objects, galaxies have the same surface brightness regardless of the distance from which they are viewed. Thus, even from a nearby spacecraft, they are not bright.

If you really want to know what a galaxy looks like up close, examine the one you're already inside. The Milky Way, which consists of our own galaxy seen edge-on from the inside, is rewarding to dark-adapted eyes, and spectacular in photographs, but is easily drowned out by city lights. Under a dark sky, it shows a prominent dust lane (the Great Rift in Cygnus and Aquila) as well as clouds of stars, especially in Scutum and Sagittarius.

All the stars in the night sky are part of our own galaxy. Its nucleus is in Sagittarius but is hidden by interstellar dust; we can observe the nucleus only as a radio source.

Our galaxy has two prominent **satellite galaxies**, the Magellanic Clouds, both too far south to see from Europe or North America. The Magellanic Clouds look like detached pieces of the Milky Way, and small telescopes easily resolve them into stars, with a scattering of clusters and nebulae. They are comparable to M32 and M110, the satellites of M31.

There are also nine or more **dwarf galaxies** in our immediate neighborhood; more are likely to be discovered. A dwarf galaxy is a faint, sparse cloud of stars, and is detectable with only the largest telescopes. The best known of these dwarfs is the "Fornax system," possibly within reach of amateur astrophotography as a cloud 1° in diameter surrounding the thirteenth-magnitude globular cluster NGC 1049. The most recent discovery is the Sagittarius dwarf galaxy, which includes one conspicuous globular cluster, M54.

11.2.5 Distant galaxies

The nearest external galaxies are M31 (with its satellites M32 and M110) and M33. Only the largest Earth-based telescopes resolve these into stars, but a large emission nebula (**H II region**) in M33 is sometimes easier to see than M33 itself; it is designated NGC 604.

Galaxies are classified as **elliptical** (round or elongated), **spiral**, **barred spiral**, or **irregular** (Figure 11.1). Between ellipticals and spirals is the **lenticular** class, for galaxies that flat like a spiral but lack spiral arms; in this book I do not distinguish them from spirals.

Relatively large amateur telescopes show the spiral arms of M51, and the irregular shape of M82 is obvious even in a 4-inch (10-cm), but the structure of most galaxies is not evident visually; they are just "faint fuzzies." Spiral arms

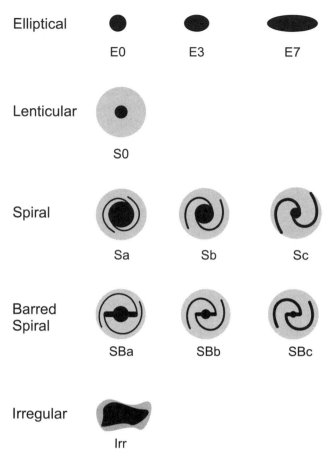

Figure 11.1. Classification of galaxies originally published by Edwin Hubble in 1936. Many variations are now in use.

normally trail behind the rotation of the galaxy, but NGC 4622 in Centaurus has two sets of arms winding in opposite directions. Note that an arm is not a solid object – it is a wave effect, somewhat like a thick spot propagating through a traffic jam, and need not move in the same direction as the stars in it.

11.2.6 Active galaxies and quasars

An active galaxy is one with an unusually bright, variable nucleus whose spectrum shows that much of the light comes from something other than ordinary stars. The light of active galactic nuclei is thought to be emitted by matter falling into a black hole.

Relatively normal-looking galaxies with active nuclei are called **Seyfert galaxies** after their discoverer, Carl Seyfert. The most conspicuous Seyfert galaxy is M77, whose nucleus varies from about magnitude 10.4 to 11.2.

More exotic are the **quasars** (quasi-stellar, i.e., starlike objects). These look like active galaxies without the galaxy; only the nucleus is visible, probably because it is immensely brighter than the stars around it. Highly variable quasars are termed **BL Lacertae objects** because one of them, BL Lacertae, was once classified as a variable star.

The spectrum of a quasar contains hydrogen lines tremendously Doppler shifted by extremely rapid motion away from the Earth. If this Doppler shift is proportional to their distance – as it is for the galaxies whose distances can be determined by other means – then quasars are immensely remote (3000 to 12 000 million light-years) and therefore extremely luminous. One of them, 3C 273, is within the reach of amateur telescopes (p. 188), and amateurs can usefully chart its variations, which range from about magnitude 12.2 to 13.0. In so doing, an amateur with a 8-inch (20-cm) telescope can touch the frontiers of the known universe.

11.3 Observing techniques

11.3.1 Star clusters

In the telescope, star clusters are rewarding because they are bright enough to be seen even near city lights, and they often look better to the eye than in photographs. The reason is that the eye can see differences in brightness that photographs conceal. Thus a bright open cluster looks, in a photograph, like a scattering of white dots, but reveals itself to the eye as a three-dimensional collection of stars differing in magnitude and coloration.

Because of their remoteness, globular clusters tend to look like fuzzy balls of cotton. A 3.5-inch (9-cm) telescope will resolve some of them into stars, such as M4; a 8-inch (20-cm) will resolve many more. In photographs, the center of a globular is an overexposed white mass, but the eye sees it as a huge sparkling ball of stars.

When viewing either type of cluster, ask yourself questions: Do the stars form lines, triangles, or other patterns? Are any of them conspicuously red? Are there any dark lanes, indicating gaps in the cluster or dust clouds in front of it? Dark lanes are also reported in a number of globulars; they do not show on photographs.

11.3.2 Bright nebulae

Planetary nebulae are compact and bright; M57 (the Ring Nebula) and NGC 3242, for instance, are visible even in suburban skies. Diffuse emission nebulae, however, are mostly faint and very hard to see. The only really bright ones are M42 (the Orion Nebula, extremely bright as nebulae go), M17 (the Swan), M8 (the Lagoon), and, for observers in the Southern Hemisphere, the η Carinae nebula (NGC 3372) and the Tarantula (NGC 2070). Reflection nebulae are invariably faint; the only one easily viewable in suburban skies is M78.

Telescopic observers of bright nebulae are rewarded by being able to see a far greater range of brightnesses than film can record. Drawing a nebula is a challenging but worthwhile enterprise; Sir John Herschel's drawings of M42 are still useful, as are the drawings in Mallas and Kreimer's *Messier Album*. One technique is to make a negative image in soft pencil on white paper, then reverse it to white on black by scanning it with a computer.

11.3.3 "Faint fuzzies"

Most galaxies and diffuse nebulae are near the limit of visibility, and the challenge is to see them at all. Here are some guidelines.

Seek clear, dark skies. Some really determined observers go to remote deserts, as much as 50 miles from the nearest human habitation. Few of us can do that, but a journey of just a few miles often brings a big payoff. Naturally, the weather must be as clear as possible.

Use medium power. Low power gives the brightest images but not, paradoxically, the best view of faint objects. At medium power, the sky background is darker and you are using the most transparent part of the lens of your eye. Many serious deep-sky observers recommend a 1.5- or 2-mm exit pupil, but always try several powers and use the one that works best.

Make the most of your night vision. Keep yourself healthy, well-fed, well-hydrated, and warm. Avoid alcohol and tobacco. Wear sunglasses during the day. For full dark adaptation, spend at least 30 minutes in the dark. Just before making a difficult observation, some observers shield their eyes from all light for five or ten minutes. A dark cloth over your head at the telescope can help by excluding light from other sources; even first-magnitude stars in the sky can sometimes be bothersome.

Use averted vision. The central region of the eye is less sensitive than the periphery, so look slightly away from the faint object you want to see. You can often make a faint object more conspicuous by moving the telescope back and forth.

Be aware of intermittent vision and "blindsight." When something is at the limit of visibility, you will not see it continuously; it will seem to fade in and out of view as you stare at it. You may experience what neurologists call "blindsight," the conviction that there is something in a particular place even though you have no conscious visual image of it.

Use filters strategically. By darkening the sky background, a nebula filter (light-pollution filter) can help bring out detail in emission nebulae. Filters help a great deal less with galaxies and other objects. Remember that a filter can never brighten objects, only darken them, so there is little need for a light-pollution filter unless the sky background is bright enough to need darkening – which it will often be, even at a country site.

11.3.4 Magnitude and surface brightness

Like stars, nebulae and galaxies have magnitudes indicating the total amount of light they send toward us, but the light of a nebula or galaxy is not concentrated in a point, and the magnitude itself is not a good indication of visibility. For example, M13 and the North America Nebula are both about 6th magnitude. M13 is clearly visible to the naked eye, but the North America Nebula is seldom seen visually because the same amount of light is spread over a much wider area.

The surface brightness of extended objects is measured as *magnitude per square arc-second* (m''), denoting the brightness that would result from spreading a star of a given magnitude over a $1'' \times 1''$ square patch of sky.

The planets are typically about $5\,m''$; the brightest nebulae, about $17\,m''$; and galaxies, typically $21\,m''$ in the center and $25\,m''$ at the edges. Because galaxies are so faint, their brightness is often given as magnitude per square arc-minute (m'), where m' and m'' are related by the simple formula:

$$m' = m'' - 8.9$$

Recall that magnitudes are logarithmic, so multiplication of light intensity corresponds to subtraction of magnitudes.

The night sky is about $19\,m''$ or $20\,m''$. How, then, do we even manage to see 21-m'' galaxies? The answer is, of course, that a 20-m'' sky plus a 21-m'' galaxy adds up to something slightly brighter than $20\,m''$. To be precise, the combination is about $19.6\,m''$, and hence distinguishable from the 20-m'' background, if only barely. (The rules for adding and subtracting m'' are the same as those for ordinary magnitudes; see p. 112.)

CCD cameras are especially good at subtracting skyglow and recovering the image of the galaxy. The human eye is not; whether with stars or with galaxies, a 0.4-magnitude difference in brightness, as in this example, is hard to see.

If an object is uniformly bright, its surface brightness can be calculated from its total magnitude m with the formula:

$$m'' = m + 2.5 \log_{10}\left(\frac{\pi}{4}D^2\right)$$

where D is the apparent diameter in arc-seconds. Most objects, however, are considerably brighter in the center and fainter at the periphery. For more about the surface brightness of galaxies, see *Sky Catalogue 2000.0*, vol. 2, pp. xxxi–xxxii.

11.4 Catalogues and designations

11.4.1 The Messier (M) catalogue

The M numbers of well-known deep-sky objects are the work of Charles Messier [pronounced *mess-YAY*] (1730–1817). While searching for comets, Messier discovered many clusters, nebulae, and galaxies. He also investigated nebulae reported by others, such as M40, which he correctly identified as a double star.

The Messier catalogue has been revised since Messier's time. The original catalogue, published in the *Connaissance des Temps* (French astronomical almanac)

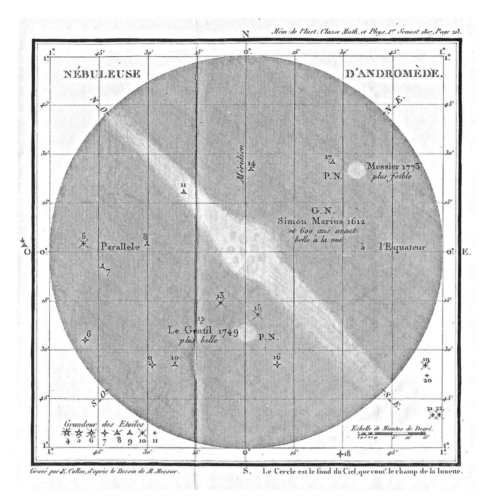

Figure 11.2. Messier's drawing of M31, M32, and M110, proving that he observed M110 even though he himself did not list it in his catalogue. From *Mémoires de la classe des sciences mathématiques et physiques de l'Institut National de France,* 1807 (1) p. 214.

for 1787, listed 103 objects; many of those after M68 were discovered by Messier's colleague Pierre Méchain [*may-SHAN*]. Messier's positions for M47, M48, and M91 were incorrect, and those of M102 and M103 were not given at all. Méchain later reported that M102 was a duplicate observation of M101. Later observers corrected the other errors.

The designations M104 to M110 were applied by modern writers to objects that Messier observed but did not catalogue. For the whole story, see John H. Mallas and Evered Kreimer, *The Messier Album* (Cambridge, 1978), which includes a copy of Messier's original catalogue in French. Two other good handbooks are K. Glyn Jones, *Messier's Nebulae and Star Clusters* (Cambridge, 1991), and S. J. O'Meara, *The Messier Objects* (Cambridge, 1998).

Table 11.1. *The Messier catalogue*

Object		Const.	Type	R.A.	Dec.	Mag.	Size
M1	NGC 1952	Tau	Supernova remnant	05^{h}34.5^m	+22°01′	8.4	6′ × 4′
M2	NGC 7089	Aqr	Globular cluster	21^{h}33.5^m	−00°49′	6.5	12.9′
M3	NGC 5272	CVn	Globular cluster	13^{h}42.2^m	+28°23′	6.2	16.2′
M4	NGC 6121	Sco	Globular cluster	16^{h}23.6^m	−26°32′	5.6	26.3′
M5	NGC 5904	Ser	Globular cluster	15^{h}18.6^m	+02°05′	5.6	17.4′
M6	NGC 6405	Sco	Open cluster	17^{h}40.1^m	−32°13′	5.3	25.0′
M7	NGC 6475	Sco	Open cluster	17^{h}53.9^m	−34°49′	4.1	80.0′
M8	NGC 6523	Sgr	Emission nebula	18^{h}03.8^m	−24°23′	6.0	90′ × 40′
M9	NGC 6333	Oph	Globular cluster	17^{h}19.2^m	−18°31′	7.7	9.3′
M10	NGC 6254	Oph	Globular cluster	16^{h}57.1^m	−04°06′	6.6	15.1′
M11	NGC 6705	Sct	Open cluster	18^{h}51.1^m	−06°16′	6.3	14.0′
M12	NGC 6218	Oph	Globular cluster	16^{h}47.2^m	−01°57′	6.7	14.5′
M13	NGC 6205	Her	Globular cluster	16^{h}41.7^m	+36°28′	5.8	16.6′
M14	NGC 6402	Oph	Globular cluster	17^{h}37.6^m	−03°15′	7.6	11.7′
M15	NGC 7078	Peg	Globular cluster	21^{h}30.0^m	+12°10′	6.2	12.3′
M16	NGC 6611	Ser	Open cluster	18^{h}18.8^m	−13°47′	6.4	7.0′
M17	NGC 6618	Sgr	Emission nebula	18^{h}20.8^m	−16°11′	7.0	11.0′
M18	NGC 6613	Sgr	Open cluster	18^{h}19.9^m	−17°08′	7.5	9.0′
M19	NGC 6273	Oph	Globular cluster	17^{h}02.6^m	−26°16′	6.8	13.5′
M20	NGC 6514	Sgr	Emission nebula	18^{h}02.6^m	−23°02′	9.0	28.0′
M21	NGC 6531	Sgr	Open cluster	18^{h}04.6^m	−22°30′	6.5	13.0′
M22	NGC 6656	Sgr	Globular cluster	18^{h}36.4^m	−23°54′	5.1	24.0′
M23	NGC 6494	Sgr	Open cluster	17^{h}56.8^m	−19°01′	6.9	27.0′
M24	—	Sgr	Milky Way patch	18^{h}16.9^m	−18°29′	4.6	90′
M25	IC 4725	Sgr	Open cluster	18^{h}31.6^m	−19°07′	6.5	40.0′
M26	NGC 6694	Sct	Open cluster	18^{h}45.2^m	−09°24′	8.0	15.0′
M27	NGC 6853	Vul	Planetary nebula	19^{h}59.6^m	+22°43′	7.4	8.0′ × 5.7′
M28	NGC 6626	Sgr	Globular cluster	18^{h}24.5^m	−24°52′	6.8	11.2′
M29	NGC 6913	Cyg	Open cluster	20^{h}23.9^m	+38°32′	7.1	7.0′
M30	NGC 7099	Cap	Globular cluster	21^{h}40.4^m	−23°11′	7.2	11.0′
M31	NGC 224	And	Spiral galaxy	00^{h}42.7^m	+41°16′	3.4	178′ × 63′
M32	NGC 221	And	Elliptical galaxy	00^{h}42.7^m	+40°52′	8.1	8′ × 6′
M33	NGC 598	Tri	Spiral galaxy	01^{h}33.9^m	+30°39′	5.7	73′ × 45′
M34	NGC 1039	Per	Open cluster	02^{h}42.0^m	+42°47′	5.5	35.0′
M35	NGC 2168	Gem	Open cluster	06^{h}08.9^m	+24°20′	5.3	28.0′
M36	NGC 1960	Aur	Open cluster	05^{h}36.1^m	+34°08′	6.3	12.0′
M37	NGC 2099	Aur	Open cluster	05^{h}52.4^m	+32°33′	6.2	24.0′
M38	NGC 1912	Aur	Open cluster	05^{h}28.4^m	+35°50′	7.4	21′
M39	NGC 7092	Cyg	Open cluster	21^{h}32.2^m	+48°26′	5.2	32′
M40	Winnecke 4	UMa	Double star	12^{h}22.4^m	+58°05′	8.4	0.8′
M41	NGC 2287	CMa	Open cluster	06^{h}46.0^m	−20°44′	4.6	38′
M42	NGC 1976	Ori	Emission nebula	05^{h}35.4^m	−05°27′	4.0	85′ × 60′
M43	NGC 1982	Ori	Emission nebula	05^{h}35.6^m	−05°16′	9.0	20′ × 15′
M44	NGC 2632	Cnc	Open cluster	08^{h}40.1^m	+19°59′	3.7	95′

Table 11.1. (*cont.*)

Object	Const.	Type	R.A.	Dec.	Mag.	Size
M45	Pleiades	Tau	Open cluster	$03^h47.0^m$	$+24°07'$	1.6	$110'$
M46	NGC 2437	Pup	Open cluster	$07^h41.8^m$	$-14°49'$	6.0	$27'$
M47	NGC 2422	Pup	Open cluster	$07^h36.6^m$	$-14°30'$	5.2	$30'$
M48	NGC 2548	Hya	Open cluster	$08^h13.8^m$	$-05°48'$	5.5	$54'$
M49	NGC 4472	Vir	Elliptical galaxy	$12^h29.8^m$	$+08°00'$	8.4	$9' \times 7.5'$
M50	NGC 2323	Mon	Open cluster	$07^h03.2^m$	$-08°20'$	6.3	$16'$
M51	NGC 5194	CVn	Spiral galaxy	$13^h29.9^m$	$+47°12'$	8.4	$11' \times 7'$
M52	NGC 7654	Cas	Open cluster	$23^h24.2^m$	$+61°35'$	7.3	$13'$
M53	NGC 5024	Com	Globular cluster	$13^h12.9^m$	$+18°10'$	7.6	$12.6'$
M54	NGC 6715	Sgr	Globular cluster	$18^h55.1^m$	$-30°29'$	7.6	$9.1'$
M55	NGC 6809	Sgr	Globular cluster	$19^h40.0^m$	$-30°58'$	6.3	$19'$
M56	NGC 6779	Lyr	Globular cluster	$19^h16.6^m$	$+30°11'$	8.3	$7.1'$
M57	NGC 6720	Lyr	Planetary nebula	$18^h53.6^m$	$+33°02'$	8.8	$1.4' \times 1.0'$
M58	NGC 4579	Vir	Spiral galaxy	$12^h37.7^m$	$+11°49'$	9.7	$5.5' \times 4.5'$
M59	NGC 4621	Vir	Elliptical galaxy	$12^h42.0^m$	$+11°39'$	9.6	$5' \times 3.5'$
M60	NGC 4649	Vir	Elliptical galaxy	$12^h43.7^m$	$+11°33'$	8.8	$7' \times 6'$
M61	NGC 4303	Vir	Spiral galaxy	$12^h21.9^m$	$+04°28'$	9.7	$6' \times 5.5'$
M62	NGC 6266	Oph	Globular cluster	$17^h01.2^m$	$-30°07'$	6.5	$14.1'$
M63	NGC 5055	CVn	Spiral galaxy	$13^h15.8^m$	$+42°02'$	8.6	$10' \times 6'$
M64	NGC 4826	Com	Spiral galaxy	$12^h56.7^m$	$+21°41'$	8.5	$9.3' \times 5.4'$
M65	NGC 3623	Leo	Spiral galaxy	$11^h18.9^m$	$+13°05'$	9.3	$8' \times 1.5'$
M66	NGC 3627	Leo	Spiral galaxy	$11^h20.2^m$	$+12°59'$	8.9	$8' \times 2.5'$
M67	NGC 2682	Cnc	Open cluster	$08^h50.4^m$	$+11°49'$	6.1	$30'$
M68	NGC 4590	Hya	Globular cluster	$12^h39.5^m$	$-26°45'$	7.8	$12'$
M69	NGC 6637	Sgr	Globular cluster	$18^h31.4^m$	$-32°21'$	7.6	$7.1'$
M70	NGC 6681	Sgr	Globular cluster	$18^h43.2^m$	$-32°18'$	7.9	$7.8'$
M71	NGC 6838	Sge	Globular cluster	$19^h53.8^m$	$+18°47'$	8.2	$7.2'$
M72	NGC 6981	Aqr	Globular cluster	$20^h53.5^m$	$-12°32'$	9.3	$5.9'$
M73	NGC 6994	Aqr	Quadruple star	$20^h58.9^m$	$-12°38'$	9.0	$2.8'$
M74	NGC 628	Psc	Spiral galaxy	$01^h36.7^m$	$+15°47'$	9.4	$10.2' \times 9.5'$
M75	NGC 6864	Sgr	Globular cluster	$20^h06.1^m$	$-21°55'$	8.5	$6'$
M76	NGC 650	Per	Planetary nebula	$01^h42.4^m$	$+51°34'$	10.1	$2.7' \times 1.8'$
M77	NGC 1068	Cet	Spiral galaxy	$02^h42.7^m$	$-00°01'$	8.9	$7' \times 6'$
M78	NGC 2068	Ori	Reflection nebula	$05^h46.7^m$	$+00°03'$	8.3	$8' \times 6'$
M79	NGC 1904	Lep	Globular cluster	$05^h24.5^m$	$-24°33'$	7.7	$8.7'$
M80	NGC 6093	Sco	Globular cluster	$16^h17.0^m$	$-22°59'$	7.3	$8.9'$
M81	NGC 3031	UMa	Spiral galaxy	$09^h55.6^m$	$+69°04'$	6.9	$21' \times 10'$
M82	NGC 3034	UMa	Irregular galaxy	$09^h55.8^m$	$+69°41'$	8.4	$9' \times 4'$
M83	NGC 5236	Hya	Spiral galaxy	$13^h37.0^m$	$-29°52'$	7.6	$11' \times 10'$
M84	NGC 4374	Vir	Spiral galaxy	$12^h25.1^m$	$+12°53'$	9.1	$5'$
M85	NGC 4382	Com	Spiral galaxy	$12^h25.4^m$	$+18°11'$	9.1	$7.1' \times 5.2'$
M86	NGC 4406	Vir	Spiral galaxy	$12^h26.2^m$	$+12°57'$	8.9	$7.5' \times 5.5'$
M87	NGC 4486	Vir	Elliptical galaxy	$12^h30.8^m$	$+12°24'$	8.6	$7'$
M88	NGC 4501	Com	Spiral galaxy	$12^h32.0^m$	$+14°25'$	9.6	$7' \times 4'$

Table 11.1. (*cont.*)

Object		Const.	Type	R.A.	Dec.	Mag.	Size
M89	NGC 4552	Vir	Elliptical galaxy	$12^h35.7^m$	$+12°33'$	9.8	$4'$
M90	NGC 4569	Vir	Spiral galaxy	$12^h36.8^m$	$+13°10'$	9.5	$9.5' \times 4.5'$
M91	NGC 4548	Com	Spiral galaxy	$12^h35.4^m$	$+14°30'$	10.2	$5.4' \times 4.4'$
M92	NGC 6341	Her	Globular cluster	$17^h17.1^m$	$+43°08'$	6.4	$11.2'$
M93	NGC 2447	Pup	Open cluster	$07^h44.6^m$	$-23°52'$	6.0	$22'$
M94	NGC 4736	CVn	Spiral galaxy	$12^h50.9^m$	$+41°07'$	8.2	$7' \times 3'$
M95	NGC 3351	Leo	Spiral galaxy	$10^h44.0^m$	$+11°42'$	9.7	$4.4' \times 3.3'$
M96	NGC 3368	Leo	Spiral galaxy	$10^h46.8^m$	$+11°49'$	9.2	$6' \times 4'$
M97	NGC 3587	UMa	Planetary nebula	$11^h14.8^m$	$+55°01'$	9.9	$3.4' \times 3.3'$
M98	NGC 4192	Com	Spiral galaxy	$12^h13.8^m$	$+14°54'$	10.1	$9.5' \times 3.2'$
M99	NGC 4254	Com	Spiral galaxy	$12^h18.8^m$	$+14°25'$	9.9	$5.4' \times 4.8'$
M100	NGC 4321	Com	Spiral galaxy	$12^h22.9^m$	$+15°49'$	9.3	$7' \times 6'$
M101	NGC 5457	UMa	Spiral galaxy	$14^h03.2^m$	$+54°21'$	7.9	$22'$
M102	= M101						
M103	NGC 581	Cas	Open cluster	$01^h33.2^m$	$+60°42'$	7.4	$6'$
M104	NGC 4594	Vir	Spiral galaxy	$12^h40.0^m$	$-11°37'$	8.0	$9' \times 4'$
M105	NGC 3379	Leo	Elliptical galaxy	$10^h47.8^m$	$+12°35'$	9.3	$2'$
M106	NGC 4258	CVn	Spiral galaxy	$12^h19.0^m$	$+47°18'$	8.4	$19' \times 8'$
M107	NGC 6171	Oph	Globular cluster	$16^h32.5^m$	$-13°03'$	7.9	$10'$
M108	NGC 3556	UMa	Spiral galaxy	$11^h11.5^m$	$+55°40'$	10.0	$8' \times 1'$
M109	NGC 3992	UMa	Spiral galaxy	$11^h57.6^m$	$+53°23'$	9.8	$7' \times 4'$
M110	NGC 205	And	Elliptical galaxy	$00^h40.4^m$	$+41°41'$	8.5	$17' \times 10'$

Note: Magnitudes and sizes of clusters, nebulae, and galaxies are inherently imprecise, and published values differ widely. Those given here are from an online catalogue published by SEDS (Students for the Exploration and Development of Space, http://www.seds.org), with minor corrections from other sources.

Messier observed through the smoky air of Paris with a variety of telescopes, none larger than 8 inches (20 cm). His 7.5- and 8-inch reflectors had mirrors of such low reflectivity that their light grasp has been compared to modern telescopes half as large. Thus, under good conditions and with some training, you can expect to see all of the Messier objects with a 3.5-inch (9-cm) telescope. At the end of March or beginning of April you can even view them all in one long night – an enterprise called a "Messier marathon," challenging because some of the objects have to be viewed in twilight.

11.4.2 The Caldwell Catalogue

After viewing all the Messier objects, what do you do next? In 1995, Patrick Caldwell-Moore (Sir Patrick Moore) proposed a list of 109 objects to observe next. (He used C for Caldwell because M for Moore was already taken.) Unlike the Messier objects, the Caldwell objects range all the way to the south pole.

Table 11.2. *The Caldwell Catalogue*

Object		Const.	Type	R.A.	Dec.	Mag.	Size
C1	NGC 188	Cep	Open cluster	$00^h44.4^m$	$+85°20'$	8.1	$14'$
C2	NGC 40	Cep	Planetary nebula	$00^h13.0^m$	$+72°32'$	12.4	$0.6'$
C3	NGC 4236	Dra	Spiral galaxy	$12^h16.7^m$	$+69°28'$	9.7	$19' \times 7'$
C4	NGC 7023	Cep	Reflection nebula	$21^h01.8^m$	$+68°12'$	7.0	$13'$
C5	IC 342	Cam	Spiral galaxy	$03^h46.8^m$	$+68°06'$	9.2	$18' \times 17'$
C6	NGC 6543	Dra	Planetary nebula	$17^h58.6^m$	$+66°38'$	8.1	$0.3'$
C7	NGC 2403	Cam	Spiral galaxy	$07^h36.9^m$	$+65°36'$	8.4	$18' \times 10'$
C8	NGC 559	Cas	Open cluster	$01^h29.5^m$	$+63°18'$	9.5	$4'$
C9	Sh2-155	Cep	Emission nebula	$22^h56.8^m$	$+62°37'$	13?	$50' \times 10'$
C10	NGC 663	Cas	Open cluster	$01^h46.0^m$	$+61°15'$	7.1	$16'$
C11	NGC 7635	Cas	Emission nebula	$23^h20.7^m$	$+61°12'$	10?	$15' \times 8'$
C12	NGC 6946	Cep	Spiral galaxy	$20^h34.8^m$	$+60°09'$	8.9	$11' \times 10'$
C13	NGC 457	Cas	Open cluster	$01^h19.1^m$	$+58°20'$	6.4	$13'$
C14	NGC 869, 884	Per	Open cluster	$02^h20.0^m$	$+57°08'$	4.3	$30' \times 30'$
C15	NGC 6826	Cyg	Planetary nebula	$19^h44.8^m$	$+50°31'$	8.8	$0.5'$
C16	NGC 7243	Lac	Open cluster	$22^h15.3^m$	$+49°53'$	6.4	$21'$
C17	NGC 147	Cas	Elliptical galaxy	$00^h33.2^m$	$+48°30'$	9.3	$13' \times 8'$
C18	NGC 185	Cas	Elliptical galaxy	$00^h39.0^m$	$+48°20'$	9.2	$12' \times 10'$
C19	IC 5146	Cyg	Emission nebula	$21^h53.5^m$	$+47°16'$	10?	$12' \times 12'$
C20	NGC 7000	Cyg	Emission nebula	$20^h58.8^m$	$+44°20'$	6?	$120' \times 100'$
C21	NGC 4449	CVn	Irregular galaxy	$12^h28.2^m$	$+44°06'$	9.4	$5' \times 4'$
C22	NGC 7662	And	Planetary nebula	$23^h25.9^m$	$+42°33'$	8.3	$0.3/2.2'$
C23	NGC 891	And	Spiral galaxy	$02^h22.6^m$	$+42°21'$	9.9	$14' \times 3'$
C24	NGC 1275	Per	Irregular galaxy	$03^h19.8^m$	$+41°31'$	11.6	$2.6' \times 2'$
C25	NGC 2419	Lyn	Globular cluster	$07^h38.1^m$	$+38°53'$	10.4	$4.1'$
C26	NGC 4244	CVn	Spiral galaxy	$12^h17.5^m$	$+37°49'$	10.2	$16' \times 2.5'$
C27	NGC 6888	Cyg	Emission nebula	$20^h12.0^m$	$+38°21'$	8?	$20' \times 10'$
C28	NGC 752	And	Open cluster	$01^h57.8^m$	$+37°41'$	5.7	$50'$
C29	NGC 5005	CVn	Spiral galaxy	$13^h10.9^m$	$+37°03'$	9.8	$5.4' \times 2'$
C30	NGC 7331	Peg	Spiral galaxy	$22^h37.1^m$	$+34°25'$	9.5	$11' \times 4'$
C31	IC 405	Aur	Emission nebula	$05^h16.2^m$	$+34°16'$	12?	$30' \times 19'$
C32	NGC 4631	CVn	Spiral galaxy	$12^h42.1^m$	$+32°32'$	9.3	$15' \times 3'$
C33	NGC 6992/5	Cyg	Supernova remnant	$20^h56.4^m$	$+31°43'$	8.0	$60' \times 8'$
C34	NGC 6960	Cyg	Supernova remnant	$20^h45.7^m$	$+30°43'$	9.0	$70' \times 6'$
C35	NGC 4889	Com	Elliptical galaxy	$13^h00.1^m$	$+27°59'$	11.4	$3' \times 2'$
C36	NGC 4559	Com	Spiral galaxy	$12^h36.0^m$	$+27°58'$	9.8	$10' \times 5'$
C37	NGC 6885	Vul	Open cluster	$20^h12.0^m$	$+26°29'$	5.9	$7'$
C38	NGC 4565	Com	Spiral galaxy	$12^h36.3^m$	$+25°59'$	9.6	$16' \times 3'$
C39	NGC 2392	Gem	Planetary nebula	$07^h29.2^m$	$+20°55'$	9.2	$0.7'$
C40	NGC 3626	Leo	Spiral galaxy	$11^h20.1^m$	$+18°21'$	10.9	$3' \times 2'$
C41	Hyades	Tau	Open cluster	$04^h27.0^m$	$+16°0.5'$	0.5	$330'$
C42	NGC 7006	Del	Globular cluster	$21^h01.5^m$	$+16°11'$	10.6	$2.8'$
C43	NGC 7814	Peg	Spiral galaxy	$00^h03.3^m$	$+16°09'$	10.5	$6' \times 2'$
C44	NGC 7479	Peg	Spiral galaxy	$23^h04.9^m$	$+12°19'$	11.0	$4' \times 3'$

Table 11.2. (*cont.*)

Object		Const.	Type	R.A.	Dec.	Mag.	Size
C45	NGC 5248	Boo	Spiral galaxy	13^{h}37.5^m	+08°53′	10.2	6′ × 4′
C46	NGC 2261	Mon	Reflection nebula	06^{h}39.2^m	+08°44′	10 (var.)	2′ × 1′
C47	NGC 6934	Del	Globular cluster	20^{h}34.2^m	+07°24′	8.9	5.9′
C48	NGC 2775	Can	Spiral galaxy	09^{h}10.3^m	+07°02′	10.3	4.5′ × 3′
C49	NGC 2237-9	Mon	Emission nebula	06^{h}32.3^m	+05°03′	6.0	80′ × 60′
C50	NGC 2244	Mon	Open cluster	06^{h}32.4^m	+04°52′	4.8	24′
C51	IC 1613	Cet	Irregular galaxy	01^{h}04.8^m	+02°07′	9.3	12′ × 11′
C52	NGC 4697	Vir	Elliptical galaxy	12^{h}48.6^m	−05°48′	9.3	6′ × 4′
C53	NGC 3115	Sex	Elliptical galaxy	10^{h}05.2^m	−07°43′	9.1	8′ × 3′
C54	NGC 2506	Mon	Open cluster	08^{h}00.2^m	−10°47′	7.6	7′
C55	NGC 7009	Aqr	Planetary nebula	21^{h}04.2^m	−11°22′	8.0	0.4′
C56	NGC 246	Cet	Planetary nebula	00^{h}47.0^m	−11°53′	10.9	3.8′
C57	NGC 6822	Sgr	Irregular galaxy	19^{h}44.9^m	−14°48	8.8	10′ × 9′
C58	NGC 2360	CMa	Open cluster	07^{h}17.8^m	−15°37′	7.2	13′
C59	NGC 3242	Hya	Planetary nebula	10^{h}24.8^m	−18°38′	7.8	0.3′
C60	NGC 4038	Crv	Spiral galaxy	12^{h}01.9^m	−18°52′	10.3	2.6′ × 2′
C61	NGC 4039	Crv	Spiral galaxy	12^{h}01.9^m	−18°53′	10.7	3′ × 2′
C62	NGC 247	Cet	Spiral galaxy	00^{h}47.1^m	−20°46′	9.1	20′ × 7′
C63	NGC 7293	Aqr	Planetary nebula	22^{h}29.6^m	−20°48′	7.3	13′
C64	NGC 2362	CMa	Open cluster	07^{h}18.8^m	−24°57′	4.1	8′
C65	NGC 253	Scl	Spiral galaxy	00^{h}47.6^m	−25°17′	7.1	25′ × 7′
C66	NGC 5694	Hya	Globular cluster	14^{h}39.6^m	−26°32′	10.2	3.6′
C67	NGC 1097	For	Spiral galaxy	02^{h}46.3^m	−30°17′	9.2	9′ × 7′
C68	NGC 6729	CrA	Emission nebula	19^{h}01.9^m	−36°57′	9.7	1.0′
C69	NGC 6302	Sco	Planetary nebula	17^{h}13.7^m	−37°06′	9.6	0.8′
C70	NGC 300	Scl	Spiral galaxy	00^{h}54.9^m	−37°41′	8.7	20′ × 13′
C71	NGC 2477	Pup	Open cluster	07^{h}52.3^m	−38°33′	5.8	27′
C72	NGC 55	Scl	Spiral galaxy	00^{h}14.9^m	−39°11′	7.9	32′ × 6′
C73	NGC 1851	Col	Globular cluster	05^{h}14.1^m	−40°03′	7.3	11′
C74	NGC 3132	Vel	Planetary nebula	10^{h}07.7^m	−40°26′	9.4	0.8′
C75	NGC 6124	Sco	Open cluster	16^{h}25.6^m	−40°40′	5.8	29′
C76	NGC 6231	Sco	Open cluster	16^{h}54.0^m	−41°48′	2.6	15′
C77	NGC 5128	Cen	Elliptical galaxy	13^{h}25.5^m	−43°01′	7.0	18′ × 14′
C78	NGC 6541	CrA	Globular cluster	18^{h}08.0^m	−43°42′	6.6	13′
C79	NGC 3201	Vel	Globular cluster	10^{h}17.6^m	−46°25′	6.7	18′
C80	NGC 5139	Cen	Globular cluster	13^{h}26.8^m	−47°29′	3.6	36′
C81	NGC 6352	Ara	Globular cluster	17^{h}25.5^m	−48°25′	8.1	7′
C82	NGC 6193	Ara	Open cluster	16^{h}41.3^m	−48°46′	5.2	15′
C83	NGC 4945	Cen	Spiral galaxy	13^{h}05.4^m	−49°28′	8.7	20′ × 4′
C84	NGC 5286	Cen	Globular cluster	13^{h}46.4^m	−51°22′	7.6	9′
C85	IC 2391	Vel	Open cluster	08^{h}40.2^m	−53°04′	2.5	50′
C86	NGC 6397	Ara	Globular cluster	17^{h}40.7^m	−53°40′	5.6	26′
C87	NGC 1261	Hor	Globular cluster	03^{h}12.3^m	−55°13′	8.4	7′

Table 11.2. (*cont.*)

Object		Const.	Type	R.A.	Decl.	Mag.	Size
C88	NGC 5823	Cir	Open cluster	$15^h05.7^m$	$-55°36'$	7.9	$10'$
C89	NGC 6087	Nor	Open cluster	$16^h18.9^m$	$-57°54'$	5.4	$12'$
C90	NGC 2867	Car	Planetary nebula	$09^h21.4^m$	$-58°19'$	9.7	$0.2'$
C91	NGC 3532	Car	Open cluster	$11^h06.4^m$	$-58°40'$	3.0	$55'$
C92	NGC 3372	Car	Emission nebula	$10^h43.8^m$	$-59°52'$	6.2	$120' \times 120'$
C93	NGC 6752	Pav	Globular cluster	$19^h10.9^m$	$-59°59'$	5.4	$20'$
C94	NGC 4755	Cru	Open cluster	$12^h53.6^m$	$-60°20'$	4.2	$10'$
C95	NGC 6025	TrA	Open cluster	$16^h03.7^m$	$-60°30'$	5.1	$12'$
C96	NGC 2516	Car	Open cluster	$07^h58.3^m$	$-60°52'$	3.8	$30'$
C97	NGC 3766	Cen	Open cluster	$11^h36.1^m$	$-61°37'$	5.3	$12'$
C98	NGC 4609	Cru	Open cluster	$12^h42.3^m$	$-62°58'$	6.9	$5'$
C99	Coal Sack	Cru	Dark nebula	$12^h53.0^m$	$-63°00'$	—	$400' \times 300'$
C100	IC 2944	Cen	Open cluster	$11^h36.6^m$	$-63°02'$	4.5	$15'$
C101	NGC 6744	Pav	Spiral galaxy	$19^h09.8^m$	$-63°51'$	8.3	$16' \times 10'$
C102	IC 2602	Car	Open cluster	$10^h43.2^m$	$-64°24'$	1.9	$50'$
C103	NGC 2070	Dor	Emission nebula	$05^h38.7^m$	$-69°06'$	8.0	$40' \times 25'$
C104	NGC 362	Tuc	Globular cluster	$01^h03.2^m$	$-70°51'$	6.6	$13'$
C105	NGC 4833	Mus	Globular cluster	$12^h59.6^m$	$-70°53'$	7.3	$14'$
C106	NGC 104	Tuc	Globular cluster	$00^h24.1^m$	$-72°05'$	4.0	$31'$
C107	NGC 6101	Aps	Globular cluster	$16^h25.8^m$	$-72°12'$	9.3	$11'$
C108	NGC 4372	Mus	Globular cluster	$12^h25.8^m$	$-72°40'$	7.8	$19'$
C109	NGC 3195	Cha	Planetary nebula	$10^h09.5^m$	$-80°52'$	11?	$0.6'$

Note: Magnitudes and sizes of clusters, nebulae, and galaxies are inherently imprecise, and published values differ widely. Those given here are from an online catalogue published by *Sky & Telescope* (http://www.skypub.com), with some corrections from David Ratledge, *Observing the Caldwell Objects* (Springer, 2000), and other sources.

The list includes spectacular southern sights such as the Eta Carinae Nebula (C92) and ω Centauri (C80) that were not visible from Messier's Paris, as well as northern sights that Messier skipped, such as the Double Cluster in Perseus (C14) and the Hyades (C41).

The objects are listed in order of declination from north to south. Thus, if you live in the Northern Hemisphere, you can start at the top of the list and proceed until you get to objects within about 5° of your southern horizon.

The Caldwell Catalogue is just a suggested observing program for deep-sky enthusiasts. Almost all of the objects were already well known by other designations. A couple of them have been criticized as excessively difficult, such as C9 (Sharpless 2-155), a very faint nebula in Cepheus, and C31 (IC 405), a nebula in Auriga. In *Observing the Caldwell Objects* (Springer, 2000), David Ratledge recommends approaching C31 with a 20-inch telescope and a narrow-band nebula filter. C9 is even harder.

11.4.3 The Herschel (H) Catalogue

Designations such as H.VI.34 or 34^6 refer to the catalogue of Sir William Herschel, the first serious deep-sky observer. In 1783, Herschel began sweeping the sky with his 18.7-inch (47-cm) Newtonian. He was not just sightseeing; besides collecting nebulae and clusters, he was investigating the shape of our galaxy, and he got it largely right. His catalogue was published in the *Philosophical Transactions* of the Royal Society (London) for 1786, 1789, and 1802.

Herschel divided objects into eight classes, described in his own words:

I. Bright nebulae
II. Faint nebulae
III. Very faint nebulae
IV. Planetary nebulae [a term Herschel invented]
V. Very large [i.e., extended, diffuse] nebulae
VI. Very compressed and rich clusters of stars
VII. Pretty much compressed clusters of large or small stars
VIII. Coarsely scattered clusters of stars

Figure 11.3 shows part of his catalogue, which inaugurated a system of abbreviations still in use. For example, Herschel's first object in the fifth class, designated H.V.1 or 1^5, is described as

cB, mE, sp nf, mbM

or, expanding the abbreviations,

considerably bright, much extended from south preceding to north following (i.e., southwest to northeast), much brighter in the middle.

This is the galaxy NGC 253, discovered by Herschel's sister Caroline. Sir William locates it 128^{m}27^s following and 1°39′ north of the star 18 Piscis Austrini, leaving it to others to compute the exact right ascension and declination.

The Herschel catalogue is of lasting value to amateur astronomers because the descriptions can be compared directly to the views in amateur telescopes. The Astronomical League (http://www.astroleague.org) has published a "Herschel 400" list and offers a certificate to amateurs who observe the 400 brightest of Herschel's 2500 objects. Unlike their Messier certificate, the Herschel certificate is valid even if you use a computerized telescope to find the objects.

11.4.4 NGC, IC, RNGC, and CNGC

The *New General Catalogue* (NGC), published by J. L. E. Dreyer in 1888, and its sequel the *Index Catalogue* (IC), is still the master list of clusters, nebulae, and galaxies used by most astronomers today.

new Nebulæ and Clusters of Stars. 493

IV.	1784	Stars.		M. S.		D.M.	Ob.	Description.
18	Oct. 6	14 Androm.	p	6 11	n	3 16	4	B. R. a planetary p. well defined disk. 15" dia^r with a 7 feet reflector.
19	16	5 Monoc.	p	7 6	f	0 10	1	A st. of the 9 magnitude, with m. chev. i elliptical.
20	— - -		p	3 42	n	0 3	1	A st. of the 11 or 12 mag. affected like the foregoing, but vF.
21	Nov. 20	12 Leporis	p	8 48	n	0 24	1	vS. stellar. vBN. and vF. chev. not quite central.
22	Dec. 9	7 (ξ) Navis	f	3 10	f	1 28	2	L. pB. R. er. 6 or 7' d. a faint red colour visible. A st. 8 mag. not far from the center, but not connected. 2d ob. 9 or 10' d.
23	1785. Jan. 6	75 Ceti	p	4 40	f	0 6	1	cB. a vBN. with a chev. of 3 or 4' d.
24	—	50 (ζ) Orio	f	0 57	f	0 17	1	A Bst. with m. chev. 5' l. 4' b.
25	31	19 Navis	p	67 0	n	1 15	1	A pst. with vF. and vS. m. chev. iF.
26	Feb. 1	34 (γ) Erid	f	16 16	n	0 49	2	vB. perfectly R. or vl. elliptical. planetary but ill defined disk. 2d obs. r. on the borders, and is probably a very compressed cluster of stars at an immense distance.
27	7	6 (3b) Crater	p	28 39	n	1 25	2	Beautiful, brilliant, planetary disk ill defined, but uniformly B. the light of the colour of Jupiter. 40" d. 2d obs. near 1' d. by estimation.
28	—	31 Crateris	f	1 0	n	0 47	1	pB. L. opening with a branch, or two nebulæ very faintly joined. The f. is smallest.
29	8	4 (1) Crateris	f	3 36	n	0 16	1	A Sst. with an eF. brush p. perceived in gaging. ver. 240.

Fifth class. Very large nebulæ.

V.	1783	Stars.		M. S.		D.M.	Ob.	Description.
1	Oct. 30	18(s) Pis. aust.	f	128 17	n	1 39	6	cB. mE. sp nf. mbM. Above 50' l. and 7 or 8' b. C. H. See note.
2	1784 Jan. 24	10 (r) Virgin	f	24 46	n	0' 17	4	cB. mE. np ff. mbM. er. 9 or 10' l with a branch towards the np.

S f f 2

Figure 11.3. A page of Sir William Herschel's catalogue of nebulae and clusters (*Philosophical Transactions*, 1786). Reproduced by permission of the President and Council of the Royal Society.

Most of the data in the NGC came from Sir John Herschel's catalogues, published in the *Philosophical Transactions* for 1833 and 1864, combining his own observations with those of his father and other astronomers. Sir John dropped the system of classes, choosing instead to number objects consecutively in order of right ascension.

Dreyer added data from more observers and renumbered the whole sequence again. The NGC covers the whole sky, and then, because it was published in two parts, the IC wraps around the whole sky twice. Because of precession, the objects are no longer in exact order of right ascension, but they are still close.

The NGC and IC continue Herschel's system of heavily abbreviated descriptions, although the descriptions are no longer all by the same observer. For instance, NGC 253 (whose description by Sir William Herschel was quoted on p. 158) is described by Dreyer as "vvB, vvL, vmE54°, gbM," that is, "very very bright, very very large, very much extended in position angle 54°, gradually brighter toward the middle." The usual scale of brightness is:

Excessively bright
Very very bright
Very bright
Bright
Considerably bright
Pretty bright
Pretty faint
Considerably faint
Faint
Very faint
Very very faint
Excessively faint

Note that "pretty" means "slightly" and "considerably" means "somewhat." The English language has changed a bit since 1786, and Sir William Herschel was not a native speaker in the first place.

The *Revised New General Catalogue* (RNGC), by J. W. Sulentic and W. G. Tifft (Arizona, 1973), adds new descriptions based on modern observatory photographs and flags many of the original objects as missing – perhaps superfluously, since photographs notoriously fail to distinguish loose star clusters from the stars in the background.

The *Computerized New General Catalogue* (CNGC) is another revision of the NGC used, as far as I can determine, only in Meade telescopes. It rates objects for visual prominence and lists some of them as "Coordinates Only," which means that the object is either nonexistent or not suitable for telescopic observation. The RNGC and CNGC use exactly the same numbers as the original NGC.

Today, the most convenient printed edition of the NGC is Roger W. Sinnott's *NGC 2000.0* (Sky Publishing, 1988), which includes the original NGC data with all of Dreyer's published corrections, plus epoch 2000 coordinates, together with a modern classification of the types of objects, distinguishing nebulae from galaxies and indicating which objects were marked as missing in the RNGC.

Still more revision is needed, because quite a few NGC and IC objects are still unidentified or disputed. Harold G. Corwin, Jr., of the California Institute of Technology, is leading a revision project whose web site, http://www.ngcic.com, contains the full contents of the catalogues (with all possible corrections and updates) and a great deal of supplementary information. Amateurs are encouraged to participate in the project.

11.4.5 Other important catalogues

The NGC and IC don't list everything. Meade LX200 telescopes also include the complete *Uppsala General Catalogue* (UGC) of over 12 000 galaxies, published in 1973 by Uppsala Astronomical Observatory in Sweden. Today's master list of galaxies is the *Catalogue of Principal Galaxies* (PGC), by G. Paturel (Lyon, 1989), which subsumes all earlier catalogues. For faint emission nebulae, an often-cited catalogue is that of S. Sharpless in the *Astrophysical Journal,* Supplement Series 4, 1959.

For planetary nebulae, PN or PLN refers to the *Strasbourg–ESO Catalogue of Galactic Planetary Nebulae* (1992), which retains the numbers of the earlier PK (Perek–Kohoutek) catalogue (1964). The numbers are based on galactic latitude and longitude (which don't change with precession) so that numbers will be permanent and objects can be inserted without disrupting the sequence. All of these catalogues, along with many others, are available online at http://adc.gsfc.nasa.gov.

There is no master catalogue of open clusters, but two lists that are often cited, besides the NGC, are those of P. J. Melotte (*Memoirs of the Royal Astronomical Society,* 1915) and P. Collinder (*Annals of the Observatory of Lund* [Sweden], 1931). All of these catalogues, along with many others, are available online at http://adc.gsfc.nasa.gov.

11.5 Handbooks, classic and modern

11.5.1 Smyth's *Cycle of Celestial Objects*

Modern observers enjoy comparing their telescopic views to those of earlier generations, especially those of the Victorian era, when astrophotography did not yet exist and eloquent verbal descriptions were highly valued.

The oldest handbook still in wide use is *A Cycle of Celestial Objects,* by Admiral W. H. Smyth (1844).[1] Its second volume, *The Bedford Catalogue,* was recently

---

[1] As best I can determine, Smyth pronounced his name with the same vowel as in *smile,* a long *i,* not like *Smith.*

reprinted by Willmann-Bell (Richmond, Virginia). That volume consists of detailed and often flowery descriptions of double stars, clusters, and nebulae as observed through a 5-inch (12.5-cm) refractor (Figure 11.4).

Smyth's double-star positions are sometimes inaccurate and were at one time falsely believed to be forged. His magnitude scale is idiosyncratic (see p. 114). His Latin, Greek, and Arabic are shaky; he always misspells the genitive *Comae Berenices* and occasionally butchers other ancient words.

But Smyth was an *amateur* in the best sense. He loved deep-sky observing and did his best to communicate his enjoyment to others and to establish amateur astronomy as a practical pursuit. We all owe him a great debt.

11.5.2 Webb's *Celestial Objects for Common Telescopes*

The next great handbook – also still in use – was the Rev. T. W. Webb's *Celestial Objects for Common Telescopes* (1859), which went through many editions and is now available from Dover Publications (New York). The reprint edition includes NGC numbers and positions for epoch 2000.

By "common telescopes" Webb meant affordable 3-inch refractors and the like; he was trying to dispel "aperture fever" and the feeling that only the largest telescopes were useful. His first volume, about the Sun, Moon, and planets, is full of vivid descriptions and historical anecdotes. The second volume is a concise catalogue of double stars, clusters, nebulae, and asterisms, with descriptions. A sample:

> [NGC] 6656 (M. 22). XVIII^h 31^m.5, S. 23°59′. Beautiful bright cl., very interesting from visibility of components, largest 10 and 11 mg., which makes it a valuable object for common telescopes, and a clue to the structure of many more distant or difficult neb. h. makes all the stars of two sizes, 11 and 15 mg., as if 'one shell over another,' and thinks the larger ones ruddy. Midway between μ and σ.

This is of course the globular cluster M22; and "h." is Sir John Herschel; and "neb." includes unresolved clusters and galaxies. The cluster is midway between the stars μ Sagittarii and σ Sagittarii. Pious to the last, Webb closes his book by quoting a Latin prayer composed by Kepler.

11.5.3 Hartung and Burnham

Observers south of the equator had to wait another century for E. J. Hartung's *Astronomical Objects for Southern Telescopes* (1968). This book also covers much of the northern sky and is therefore of worldwide interest. It gives visual descriptions in the spirit of Webb and Smyth, though illuminated by modern knowledge. A revised edition was published by Cambridge University Press in 1996.

CCLXIX. 2 ♅. VI. GEMINORUM.

Æ 6ʰ 45ᵐ 56ˢ   Prec. + 3ˢ·50
Dec. N 18° 10'·5   —— S 3"·99

Mean Epoch of the Observation . . . . 1837·91

A compressed cluster, on the calf of Pollux's right leg, one-third of the distance from Pollux to Rigel, on a line carried from the former

star between the second and third " bullions" of Orion's belt to the latter: discovered by ♅. in 1783, and forming No. 415 of his son's Catalogue. It is a faint angular-shaped group of extremely small stars—say 12 to 16 magnitudes—which only under the most favourable circumstances can I discern with satisfaction. The region around is immensely rich, and not at all wanting in double stars. Differentiated with γ Geminorum for a mean place; and when best seen, it is something like the hasty sketch herewith given.

CCLXX. π² CANIS MAJORIS.

Æ 6ʰ 48ᵐ 08ˢ   Prec. + 2ˢ·51
Dec. S 20° 12'·4   —— S 4"·18

Position A B 149°·0 (w 5)	Distance 45"·0 (w 5)	
—— A C 182°·5 (w 3)	—— 52"·5 (w 3)	Epoch 1834·14
—— A D 185°·0 (w 3)	—— 125"·0 (w 1)	

A coarse quadruple star, on the chest of Canis Major; where it is the middle one of three small stars, about $4\frac{1}{2}$° to the south-south-east of Sirius. A 6, flushed white; B $9\frac{1}{2}$, ruddy; C 10, ruddy; D 11, dusky. A and B were classed as a double star, 65 ♅. v.; and Piazzi, note 222 Hora VI., says, "Binæ sequuntur 10ᵃ magn. 1" circiter ad austrum." Herschel's measures were:

Pos. 154° 12' Dist. 44"·93 Ep. 1782·17

When Sir James South examined this object, he included the two companions in the *sp* quadrant, and registered it quadruple, thus:

A B Pos. 147° 57' Dist. 45"·03 Ep. 1825·04
A C   184° 18'   52"·96   1825·07
A D   185° 16'   128"·36   1825·10

On weighing all these results, there seems to have been some error in ♅.'s angle, at the first epoch.

Figure 11.4. Admiral Smyth describes H.VI.2 (NGC 2304) and π^2 Canis Minoris. Note the H-shaped monogram of Sir William Herschel, which is nowadays used as a symbol for Uranus.

The last of the great classics is *Burnham's Celestial Handbook,* by the late Robert Burnham, a diligent staff member of Lowell Observatory who compiled, in three volumes, all that was known about stars, clusters, and nebulae up to the late 1960s.[2] The second edition has been available since 1978 from Dover Publications and is so useful that I own two and a half copies. The information is somewhat dated but still extremely useful. The coordinates are for epoch 1950.

11.5.4 Modern handbooks

The current handbooks that I find most useful are the *Observing Handbook and Catalogue of Deep-Sky Objects,* by Christian B. Luginbuhl and Brian A. Skiff (Cambridge, second edition, 1999), and the *Night Sky Observer's Guide,* by B. R. Kepple and G. W. Sanner (Willmann-Bell, 1998). Each of these describes thousands of deep-sky objects as seen in amateur telescopes and gives up-to-date reference information and finder charts.

Also noteworthy is the *Webb Society Deep-Sky Observer's Handbook* (7 vols., Enslow Publishers, 1975–1987), a comprehensive survey of the sky carried out by amateurs using the best available equipment and techniques. Besides reference data, these books give visual descriptions and drawings of the view of each object through telescopes of various sizes. The Webb Society is still thriving and invites deep-sky observers to join; it can be reached at http://www.webbsociety.freeserve.co.uk or c/o R. W. Argyle, Institute of Astronomy, Madingley Road, Cambridge CB3 0HA, England.

Work of the same type is continuing, and new books are published frequently. Anyone can publish a list of favorite deep-sky objects – and every few months, someone does. Even your humble author has tried his hand at it. To see the results, turn the page.

---

[2] Not to be confused with another Robert Burnham who was at one time editor of *Astronomy* magazine.

Part II

200 interesting stars and deep-sky objects

Chapter 12
How these objects were chosen

Back in the 1850s, T. W. Webb noticed that amateur astronomy had become competitive – his contemporaries were trying to outdo each other by building larger and larger reflectors. He decided to break out of this trend and write a book for owners of 3-inch refractors and the like. His book, *Celestial Objects for Common Telescopes,* has become a classic and is still available.

In a similar spirit, I want to break away from the present-day tendency to equate deep-sky observing with star-hopping to the faintest possible nebulae and galaxies, which are visible only under extremely dark skies. There's more to the stellar universe than just "faint fuzzies."

The objects in this list are visible with an 8-inch (20-cm) telescope under suburban skies with a naked-eye magnitude limit of 5. Most of them are fine sights even with considerably smaller telescopes under worse conditions. Many of them are *stars* (variable, double, multiple, or unusually colored) or bright star clusters.

This list is based on my own observations. It is organized by zones of right ascension, and within each zone, in sequence from north to south. Chapter titles such as "The January–February sky" refer to the time of year when the objects are highest at 10 or 11 p.m. local time. Many of the objects, especially the more northerly ones, can be viewed for much longer than just the specified two-month period. Those in the extreme south, however, are in the sky only briefly.

Each chapter begins with a chart of the whole sky to help you identify alignment stars. On the map, constellations near the horizon are unavoidably distorted due to the map projection used.

The entry for each object gives its name(s), catalogue designations (including LX200 and NexStar star numbers, where applicable), right ascension, declination, magnitude, and size. For multiple stars, the positions and catalogue numbers are for the brightest component. For planetary nebulae, the diameter is that of the bright central region, not the periphery. All positions are for epoch 2000.0. Double-star parameters are from WDS (see p. 13). The other information is from the reference books cited elsewhere in this book, and from the following sources, which are cited by author's name:

Karkoschka, Erich. *The Observer's Sky Atlas*, 2nd edition. Springer-Verlag, 1999. (Mixed-scale atlas with very concise but useful object lists.)

Mullaney, James. "My 10 favorite deep-sky wonders." *Sky & Telescope,* December, 2000, 121–124.

Sinnott, Roger W. "Hunting for equilateral triple stars." *Sky & Telescope,* March, 1999, 100–101.

Skiff, Brian A. "Carbon stars: reddest of the red." *Sky & Telescope,* May, 1998, 90–95 (includes finder charts).

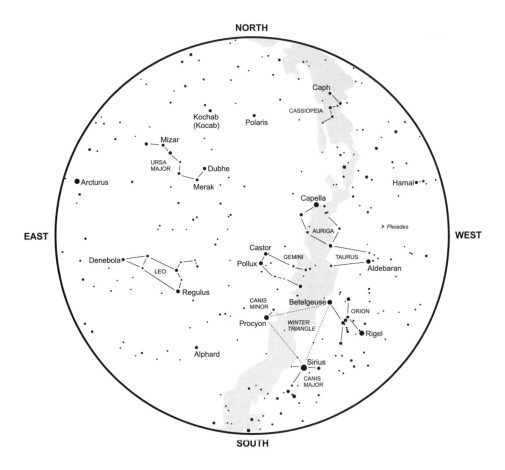

NORTH

Caph
CASSIOPEIA
Kochab
(Kocab)
Polaris
Mizar
URSA
MAJOR
Dubhe
Hamal
Arcturus
Merak
Capella
AURIGA
☆ Pleiades
EAST
WEST
Castor
GEMINI
TAURUS
Denebola
Pollux
Aldebaran
LEO
Regulus
CANIS
MINOR
Betelgeuse
ORION
Procyon
WINTER
TRIANGLE
Rigel
Alphard
Sirius
CANIS
MAJOR

SOUTH

THE SKY AT SIDEREAL TIME 8:00

Mid-January	Midnight local time
Mid-February	10 p.m. local time
Mid-March	8 p.m. local time

As seen from latitude 40° north.

British and Canadian observers will see more stars above the north horizon;
southern U.S. observers will see more stars above the south horizon.
Based on a chart created with *Starry Night Pro* astronomy software
(http://www.starrynight.com), reproduced by permission.

Chapter 13
The January–February sky (R.A. 6^h–10^h)

1   **M82**
NGC 3034

IRREGULAR GALAXY IN URSA MAJOR
9^{h}55.9^m +69°41′
Magnitude 8.4
Size 11′ × 4′

M82 is the finest irregular galaxy in the sky; its turbulent shape is evident even at low powers. M81 is almost in the same low-power field.

---

2   **M81**
NGC 3031

SPIRAL GALAXY IN URSA MAJOR
9^{h}55.6^m +69°04′
Magnitude 6.9
Size 27′ × 14′

Prominent object, though located in a relatively blank field. This rather bright galaxy is actually a spiral, though it looks like an elliptical galaxy in most telescopes. Contrast its smooth shape and texture with the irregular galaxy M82. M81 has a bright core; M82 does not.

Two 10th-magnitude stars are south of M81, and two 12th-magnitude stars are right in front of it, just south–southeast of the nucleus; they look like supernovae in M81 but are actually part of our own galaxy.

M81 is a good choice if you want to observe a galaxy under adverse conditions. It is plainly seen even in a 3.5-inch (9-cm) telescope under a town sky. (M82 is considerably fainter.) If conditions are good, have a look at the fainter galaxy NGC 3077, three quarters of a degree to the southeast.

---

3 **Caldwell 7** SPIRAL GALAXY IN CAMELOPARDALIS
NGC 2403 7^{h}37.0^m +65°36'
Magnitude 8.5
Size 18' × 10'

This is the brightest galaxy north of the celestial equator that did not receive a Messier number, according to D. F. Brocchi (*Sky & Telescope,* February 1994), who also remarks that "Camelopardalis is the absence of a constellation" – it is a large area of sky with no bright stars.

---

4 **W Ursae Majoris** ECLIPSING BINARY STAR IN URSA MAJOR
GCVS 830006 9^{h}43.8^m +55°57'
SAO 27364 Magnitude 7.7–8.5
HIP 47727 Period 8.0 hours

This dwarf eclipsing binary system consists of two Sun-like stars orbiting extremely close together, so that they are elongated into ellipsoids that nearly touch. There are two eclipses during each revolution, and both minima are equally deep; thus you can see the star dip to magnitude 8.5 every four hours. The brightening and fading are gradual because each star is nearly always partly in front of the other one.

If you watch W Ursae Majoris periodically for two or three hours, you are sure to see action. It is just 21' east of the 6.5-magnitude star SAO 27340 (NexStar Star 7613), and to the north and south are two 8.9-magnitude stars in a double equilateral triangle. About 23' west of SAO 27340 is a fine tenth-magnitude triple.

---

5 **Castor** DOUBLE STAR IN GEMINI
α Geminorum 7^{h}34.6^m +31°53'
SAO 60198 Magnitude 2.0, 2.9
HIP 36850 Separation ≈4.5"
LX200 Star 78 Position angle ≈60°
NexStar Star 1865

Castor, alias Alpha Geminorum, is a magnificent double star in 3-inch or larger telescopes. The separation is slowly increasing (see Figure 9.4 on p. 126). Recommended power, 100× to 200×.

The third component of the system, Castor C, is 73" away in position angle 164°. It is 9th magnitude and easy to see once your attention is called to it. Spectroscopy has shown that Castor C consists of a close pair of red dwarfs orbiting each other every 19.5 hours.

---

6   *ι* **Cancri**   DOUBLE STAR IN CANCER
SAO 80416   8^{h}46.7^m +28°46′
HIP 43103   Magnitudes 4.0, 6.6
NexStar Star 2216   Separation 30.4″
                                       Position angle 307°

Prominent object. Iota Cancri is a fine wide double star resembling Albireo but with more muted colors, cream-yellow and slate-blue at 40×.

---

7   **M35**   OPEN CLUSTER IN GEMINI
NGC 2168   6^{h}08.9^m +24°20′
                                       Magnitude 5.1
                                       Diameter 25′

Prominent object. This fine open cluster at the feet of Gemini fills the low-power field. Note how the stars seem to be arranged in long curving chains.

Half a degree to the southwest, just outside the field at 40×, is a much fainter and more distant cluster, NGC 2158, visible in a 4-inch (10-cm) telescope under good conditions. In the 8-inch (20-cm) under a town sky, I see it only as a fuzzy glow.

---

8   **Clown Face Nebula**   PLANETARY NEBULA IN GEMINI
Eskimo Nebula   7^{h}29.2^m +20°55′
NGC 2392   Magnitude 8.6
Caldwell 39   Diameter 15″

In observatory photographs, this nebula resembles a smiling face surrounded by a fuzzy hood. The central bright region, 15″ in diameter, surrounds a clearly visible 11th-magnitude central star. Definitely nonstellar in 5-inch (12.5-cm) at 40×; bright enough to be visible even under adverse conditions.

Right next to NGC 2392 is a star of magnitude 10.5. With averted vision, the star and the nebula are of equal brightness; when stared at directly, the star is brighter than the nebula.

At high magnification (100× or higher), the nebula shows an interesting structure, with a central star, a bright central area, and a fainter periphery 45″ in diameter.

---

9   **Beehive Cluster**   OPEN CLUSTER IN CANCER
       Praesepe   8^{h}39.9^m +19°33′
       M44   Magnitude 3
       NGC 2362   Diameter 1.2″

Prominent object. This star cluster is far too big to fit into the field of the telescope; the coordinates given are those of SAO 98013, a fine double star within it (magnitudes 7.5 and 7.8, separation 45″, position angle 157°). The cluster contains many other doubles.

The Latin name *Praesepe* (*pree-SEE-pee*) means "manger," and the stars flanking the cluster are called *Asellus Borealis* and *Asellus Australis,* the Northern and Southern Donkeys.

---

10   **20 Geminorum**   DOUBLE STAR IN GEMINI
        SAO 95795   6^{h}32.3^m +17°47′
        HIP 31158   Magnitudes 6.3, 7.0
                                                            Separation 20.0″
                                                          Position angle 211°

A nice double star, plainly double at 50× in the ETX-90, but higher power is desirable; at the corner of an almost equilateral triangle. In larger telescopes, it is a fine sight. At higher power, the brighter member looks warmer-colored.

---

11   **ζ Cancri**   TRIPLE STAR IN CANCER
        SAO 97645   8^{h}12.2^m +17°39′
        HIP 40167   Magnitudes 5.3, 6.2, 5.8
        LX200 Star 293   Separations 0.9″, 6.4″ (2000)
        NexStar Star 2056   Position angles 83°, 72° (2000)

The primary star of Zeta Cancri has two 6th-magnitude companions, one of them easily visible at 6.4″ and the other one much closer, at 0.9″, widening to 1.1″ by 2010.

Under average conditions with an 8-inch (20-cm), I saw this system as a close double at 40×, and at 250×, the brighter component was elongated. In steady air, the closer pair is cleanly split, and a 5-inch (12.5-cm) at 300× shows it as two diffraction disks partly overlapping.

The orbital period of the two closest stars is only 60 years. They were aligned east–west in 1999 and will be aligned north–south in 2020 (Figure 13.1).

ζ Cancri AB (WDS)

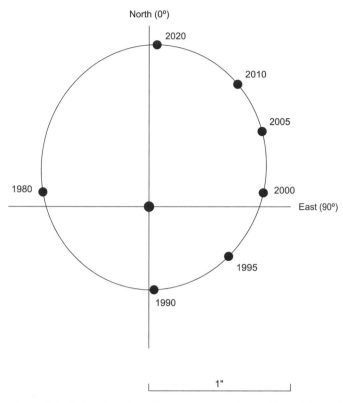

Figure 13.1. Orbit of the two closer components of ζ Cancri, flipped to match the view through a telescope with a diagonal.

12   X Cancri   SEMIREGULAR VARIABLE STAR IN CANCER
GCVS 120007 $8^h55.4^m +17°14'$
SAO 98230   Magnitude 6.0–6.5
HIP 43811
NexStar Star 2254

With a color index of 3.8 (Hipparcos), this semiregular variable is one of the redder stars in the sky, but not as red as V Aquilae (p. 216) or U Camelopardalis (p. 235). Note that the designation is X, not χ (chi).

---

13   Christmas Tree Cluster   OPEN CLUSTER WITH NEBULA IN MONOCEROS
NGC 2264 $6^h41.0^m +9°54'$
                                                    Magnitude 3.5
                                                    Diameter 30'

Prominent object. This attractive star cluster is near the Cone Nebula (also part of NGC 2264) in the constellation Monoceros. The nebula itself is large (3°) and quite faint.

The coordinates given are those of the brightest star in the cluster, S Monocerotis, magnitude 4.6, slightly variable. It is a very blue-white star of spectral type O7 and is double, with a 7.8-magnitude companion at 2.9″, position angle 213°.

The cluster is a fine sight at 40× even in a small telescope. The bright double star is resolved at 250× in an 8-inch (20-cm).

---

14   **Hubble's Variable Nebula**   REFLECTION NEBULA IN MONOCEROS
NGC 2261   6^{h}39.2^m +8°44′
Caldwell 46   Magnitude 10.5–13
                                         Size 2′ × 1′

Hubble's Variable Nebula is a compact, comet-shaped, relatively bright nebula surrounding the variable star R Monocerotis (GCVS 550001, mag. 11–14), a hot, newly formed star whose brightness has not yet stabilized. Although somewhat irregular, R Monocerotis covers its whole range from magnitudes 10.5 to 13 in a rough 350-day cycle, with some minima deeper than others. (To check on its current behavior, go to http://www.aavso.org.) The nebula varies with the star; at minimum it is rather difficult even in an 8-inch (20-cm) telescope.

---

15   **Σ 939**   TRIPLE STAR IN MONOCEROS
SAO 114112   6^{h}35.9^m +5°19′
HIP 31513   Magnitudes 8.4, 9.2, 9.4
                                         Separations 30.1″, 39.7″
                                         Position angles 106°, 50°

Striking triple star forming a nearly equilateral triangle, discovered by F. G. W. Struve. Not in telescopes' built-in catalogues; find it by right ascension and declination (in full: 06^{h}35^{m}55^s +05°18′35″). For accuracy, sync on a nearby star first.

This triple is well seen at 40×, though 80× is better; not bright, but plainly visible in a 5-inch (12.5-cm) in moonlight. It is striking in the 8-inch (20-cm) at 140×. For a diagram of this system see Figure 9.3, p. 126.

Σ stands for F. G. W. Struve, who catalogued double stars about a century ago.

---

16   **Caldwell 50**   OPEN CLUSTER IN MONOCEROS
         NGC 2244   6^{h}32.3^m +4°51′
                                                          Magnitude 4.8
                                                          Diameter 24′

This bright open cluster forms the center of the Rosette Nebula (NGC 2237–9). The nebula's glow is visible only under clear, dark skies, but you may be able to observe it as a *dark* nebula, i.e., a star-poor region encircling the cluster, where light from more distant stars is blocked by interstellar gas.

The coordinates given are of the star 12 Monocerotis, apparently the brightest star in the cluster but actually a foreground object.

---

17   **ε Monocerotis**   DOUBLE STAR IN MONOCEROS
         SAO 113810   6^{h}23.8^m +4°36′
         HIP 30419   Magnitudes 4.4, 6.6
         NexStar Star 1483   Separation 12.4″
                                                       Position angle 29°

Epsilon Monocerotis is a fine moderately close double for telescopes of all sizes, well seen at 65×, much better seen at 250×.

---

18   **NGC 2301**   OPEN CLUSTER IN MONOCEROS
                                                        6^{h}51.8^m +0°28′
                                                          Magnitude 6.0
                                                          Diameter 6′

Prominent object. A cluster with a "conspicuous chains of stars" (Karkoschka) – more precisely, a chain of stars with another chain of stars perpendicular to it. The brighter chain is visible even in the 8 × 50 finder. Rarely observed, but well worth a look.

---

19   **M48**   OPEN CLUSTER IN HYDRA
         NGC 2548   8^{h}13.7^m −5°48′
                                                          Magnitude 5.8
                                                          Diameter 30′

Prominent object. This fine cluster fills the low-power field. Its overall shape is triangular or heart-shaped.

It is, however, not entirely clear that this is actually the object Messier meant to list as M48, since the position he gave was 5° farther north.

You may find it interesting to explore and see if anything else in the vicinity fits Messier's description: "a cluster of very small (faint) stars without nebulosity." Note that he says nothing about the richness of the cluster. Precessed to epoch 2000, the position he gave is 8^{h}13.9^m −1°56′.

---

20 β **Monocerotis**   TRIPLE STAR IN MONOCEROS
SAO 133317   6^{h}28.8^m −7°02′
HIP 30867   Magnitudes 4.6, 5.0, 6.1
LX200 Star 287   Separations (from B) 7.2″, 2.9″
NexStar Star 1514   Position angles (from B) 213°, 108°

Prominent object. Beta Monocerotis is a fine triple star, well seen at 75× or higher, even in modest telescopes, when the air is steady. It was discovered by Sir William Herschel in 1781 and is a true ternary system, all three stars orbiting around their common center of gravity.

This is one of the few triple stars with all three members nearly the same brightness. At lower power (40×), it looks like a close double. High power (150× or more) displays its full glory.

Because components B and C are so close together, I give the separations and position angles from component B (magnitude 5.0) rather than component A.

---

21   **Ghost of Neptune**   PLANETARY NEBULA IN CANIS MAJOR
IC 2165   6^{h}21.7^m −12°59′
                                                                           Magnitude 10
                                                                            Diameter 4″

A small planetary nebula with high surface brightness (J. McNeil, *Sky & Telescope*, January 1999, p. 100). Use medium to high power, as its diameter is only 4″. At 140×, in the 8-inch (20-cm), it is reasonably bright and non-stellar. I call it the Ghost of Neptune because, like Neptune, it is distinguished from the surrounding stars as much by its color as by its size.

---

22 μ **Canis Majoris**   DOUBLE STAR IN CANIS MAJOR
SAO 152123   6^{h}56.1^m −14°03′
HIP 33345   Magnitudes 5.3, 7.1
NexStar Star 1655   Separation 2.8″
                                                                   Position angle 343°

Mu Canis Majoris is a close double star with color contrast; it requires steady air and power > 140×. One can imagine the stars to be red and green, though pale red and slate-blue is perhaps a better description.

---

23 **M47**
NGC 2422

OPEN CLUSTER IN PUPPIS
7^{h}36.6^m −14°29′
Magnitude 6
Diameter 25′

Prominent object. For a long time this cluster was unidentified, since Messier gave an erroneous position for it, and the object at that position was designated NGC 2478 even though it seemed not to exist. Near the center is the fine double star Σ 1121 (mags. 7.0 and 7.3, 7.8″, position angle 305°). The open cluster NGC 2423 is 40′ to the north.

24 **M46**
NGC 2437

OPEN CLUSTER IN PUPPIS
7^{h}41.8^m −14°49′
Magnitude 6.1
Diameter 20′

This distant cluster of faint stars contrasts dramatically with M47 next door. Because the stars are faint, M46 is hard to see in moonlight, though M47 is easy.

M46 contains an 11th-magnitude planetary nebula, NGC 2438. (It is not clear whether the nebula is actually part of the cluster or whether it is in the foreground or background.) At 75× to 150×, the nebula looks like a small, faint, ghostly patch on the northeast side of the cluster, easily mistaken for a thick spot in the cluster itself. A nebula filter makes it much more prominent. Its surface brightness is unusually low for a planetary nebula.

25 **NGC 2440**

PLANETARY NEBULA IN PUPPIS
7^{h}41.9^m −18°12′
Magnitude 9
Diameter 14″

This small, bright planetary nebula is much brighter than the magnitude of 11 listed in most catalogues; I estimate it as 9.0, slightly fainter than the 8.5-magnitude reddish star just 3′ to the east. The nebula is unusually symmetrical; it looks like a bright, unresolved globular cluster.

The same medium-power field also contains two triple stars, one close and one wide, and a scattering of other stars.

26   **M41**   OPEN CLUSTER IN CANIS MAJOR
NGC 2287   6^{h}46.0^m −20°45′
                                                     Magnitude 4.5
                                                     Diameter 40′

Prominent object. This bright, field-filling star cluster is a fine sight even under very adverse conditions. Under a dark country sky, it is visible to the naked eye and was noted by Aristotle over 300 years before Christ. It is easy to find 4° due south of Sirius.

---

27   **145 Canis Majoris**   DOUBLE STAR IN CANIS MAJOR
SAO 173349   7^{h}16.6^m −23°19′
HIP 35210   Magnitudes 5.0, 5.8
NexStar Star 1765   Separation 26.8″
                                                     Position angle 52°

Prominent object. Appearances can be deceiving. This fine double star, with a nice color contrast, is not a double star at all – according to Hipparcos, the components are respectively 6000 and 250 light-years away from us. Also designated h 3945 (Herschel 3945); Mullaney calls it the **Winter Albireo**. A fine sight at 40× in the NexStar 5.

---

28   **τ Canis Majoris Cluster**   OPEN CLUSTER IN CANIS MAJOR
NGC 2362   7^{h}17.8^m −24°57′
Caldwell 64   Magnitude 4.1
                                                     Diameter 8′

Prominent object. Attractive cluster surrounding the star Tau Canis Majoris, which has been called the **Mexican Jumping Star** (by Bill Arnett in 1996) because of an amusing optical illusion. Apparently, when your telescope shakes in the wind, the bright star and cluster will appear to move in different directions because the star is brighter and produces a stronger afterimage. I did not see this illusion with an 8-inch (20-cm) at 40×.

---

29   **NGC 2467**   EMISSION NEBULA IN PUPPIS
                                                     7^{h}52.6^m −26°23′
                                                     Magnitude 7
                                                     Diameter 15′

In my 8-inch (20-cm) telescope at 65× under a town sky, this unusual nebula looks like a roundish cloud about a quarter of a degree in diameter surrounding a star, although under better conditions it turns out to be irregular in shape with several stars involved. A nebula filter makes it much more prominent. Use medium rather than low power.

---

30   NGC 2451   OPEN CLUSTER IN PUPPIS
                                                       7^{h}45.4^m −37°58′
                                                       Magnitude 2.8
                                                       Diameter 1°

Prominent object. A bright cluster with "few stars, but very bright and colored ones" (Karkoschka). More precisely, a bright red star flanked by six other stars in an arrangement making two parallelograms, with a few other stars in the field. A fine object for small telescopes.

---

31   AI Velorum   SHORT-PERIOD VARIABLE STAR IN VELA
        GCVS 850063   8^{h}14.1^m −44°35′
        SAO 219640   Magnitude 6.2–6.8
        HIP 40330   Period 2 hours 41 minutes
        NexStar Star 7301

This low-amplitude pulsating variable star (δ Scuti type) has an unusually short period. Compare it to the magnitude 7.4 star 10′ to the south – and come back every half hour and compare it again. The amplitude of variation is not the same on every cycle.

---

32 γ Velorum   MULTIPLE STAR IN VELA
        SAO 219504   8^{h}09.5^m −47°20′
        HIP 39953
        LX200 Star 85
        NexStar Star 2040

Gamma Velorum is a very fine quadruple star at 40×, though located so far south that you have to go as far south as Florida to see it well. The four stars are all at different distances; what we are seeing is purely an optical alignment. The name **Regor** was bestowed on this star by Roger Chaffee as a practical joke (see p. 91).

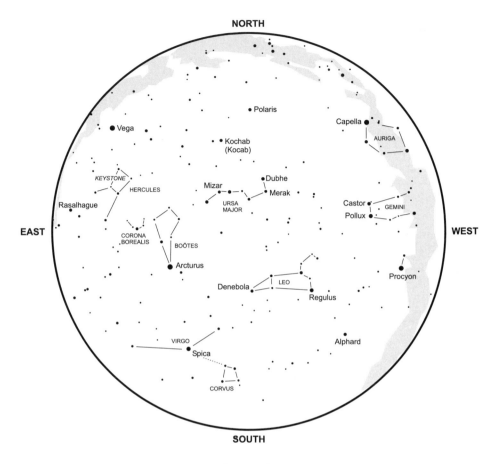

THE SKY AT SIDEREAL TIME 12:00

Mid-March	Midnight local time
Mid-April	10 p.m. local time
Mid-May	9 p.m. daylight saving time

As seen from latitude 40° north.

British and Canadian observers will see more stars above the north horizon;
southern U.S. observers will see more stars above the south horizon.

Based on a chart created with *Starry Night Pro* astronomy software
(http://www.starrynight.com), reproduced by permission.

Chapter 14
The March–April sky (R.A. 10^h–14^h)

33   **IC 3568**   PLANETARY NEBULA IN CAMELOPARDALIS

12^{h}32.9^m +82°33′

Magnitude 10

Diameter 6″

One of the brightest in a list of neglected planetaries (*Sky & Telescope*, January 1999, p. 126). Requires medium power to be clearly recognizable. In a 5-inch (12.5-cm) at 90×, it is clearly visible as a star tightly wrapped in a compact ball of fuzz, contrasting with a 10th-magnitude star 7′ to the east that is free of nebulosity.

Because it is so close to the north celestial pole, this object is more easily found in altazimuth than in equatorial mode.

---

34   **M40**   DOUBLE STAR IN URSA MAJOR
        Winnecke 4

12^{h}22.2^m +58°05′

Magnitudes 9.7, 10.2

Separation 52″

Position angle 78°

This is the only Messier object that is a double star. Messier listed it because Hevelius in 1660 had reported a nebula in this position; Messier concluded it was merely a double star blurred by Hevelius' imperfect instrument.

Note that the stars are only 10th magnitude; they are clearly visible in a 3.5-inch (9-cm) telescope but do not stand out. A considerably brighter star is in the field.

---

35   **Mizar and Alcor**   TRIPLE STAR IN URSA MAJOR
           $\zeta^{1,2}$ Ursae Majoris   13^{h}23.9^m +54°56′
           SAO 28737   Magnitudes 2.2, 3.9, 4.0
           HIP 65378   Separations 14.6″, 708″
           LX200 Star 137   Position angles 153°, 71°
           NexStar Star 3151

Prominent objects. Mizar and Alcor, Zeta-1 and Zeta-2 Ursae Majoris, are a naked-eye pair known since ancient times. Mizar is itself the first telescopic double ever discovered (by Riccioli in 1650). Between them is an unrelated star, sometimes called **Sidus Ludovicianum** ("Ludwig's Star"), that was briefly mistaken for a new planet in 1723 by a German astronomer who hastened to name it after Ludwig V of Hesse-Darmstadt.

---

36   **Whirlpool Galaxy**   SPIRAL GALAXY IN CANES VENATICI
           M51   13^{h}29.9^m +47°12′
           NGC 5194   Magnitude 8.4
                                                         Size 11′ × 7′

This galaxy shows spiral structure more prominently than any other; an 8-inch (20-cm) telescope will show the spiral arms under good conditions. A satellite galaxy at the end of one of the spiral arms is designated NGC 5195.

In surface brightness, M51 is considerably fainter than, for instance, M81 and M82; it requires a dark sky.

---

37   **Y Canum Venaticorum**   CARBON STAR IN CANES VENATICI
           GCVS 130008   12^{h}45.1^m +45°26′
           SAO 44317   Magnitude 5.0–7.5
           HIP 62223
           NexStar Star 3025

This semiregular variable is a carbon star (spectral class C5) and is a beautiful reddish color, with a color index of 2.5. The name **La Superba**, given by Secchi, refers to the dramatic appearance of its spectrum in a spectroscope. The star is not as red as, for example, X Cancri.

---

38   **Cor Caroli**   DOUBLE STAR IN CANES VENATICI
        α Canum Venaticorum   12ʰ56.0ᵐ +38°19′
        SAO 63257   Magnitudes 2.8, 5.5
        HIP 63125   Separation 19.8″
        LX200 Star 133   Position angle 230°
        NexStar Star 3071

Prominent object. This handsome double star (see Figure 9.1) is a fine sight at 40×. The name *Cor Caroli* (heart of Charles) honors Charles I of England. Some observers see a color contrast; others do not. To me, both stars look blue-white, with the companion somewhat bluer than the primary. There is another double star half a degree to the east.

---

39   **ξ Ursae Majoris**   DOUBLE STAR IN URSA MAJOR
        SAO 62484   11ʰ18.2ᵐ +31°32′
        HIP 55203   Magnitudes 4.3, 4.8
        LX200 Star 297   Separation ≈3″
        NexStar Star 2745

Xi Ursae Majoris is sometimes called **Alula Australis**. This double star has a very fast orbit seen face-on, so the position angle is changing rapidly but the separation is not (Figure 14.1). The two stars were aligned east–west in 2001,

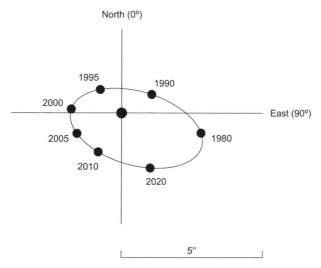

Figure 14.1. The orbit of ξ Ursae Majoris, flipped to match the view in a telescope with a diagonal.

and will be northeast–southwest in 2008, and north–south around 2015. Steady air and high power are required.

With a 5-inch (12.5-cm) at 150×, in steady air, in 2000, I saw this star as two diffraction disks separated by a dark lane, like a cell undergoing mitosis. A 9-cm shows two disks in contact; an 8-inch (20-cm) separates the stars cleanly. Both stars are somewhat yellowish.

Because of its high proper motion (over 0.6″/year), which led to inconsistent position data, this star was accidentally omitted from some editions of the Hipparcos catalogue and from charts and software prepared from Hipparcos data. For example, *TheSky* version 5 labels it but does not plot it.

---

40 **M3** GLOBULAR CLUSTER IN CANES VENATICI
NGC 5272 13^h42.2^m +28°22′
Magnitude 5.9
Diameter 19′

Prominent object. "Although M3 lies in a relatively star-poor field, it is contained in an acute triangle of 9th-magnitude stars, and an obvious topaz star borders the cluster to its northwest" (O'Meara). The orange 6th-magnitude star SAO 82944 is just outside the field.

In a 3.5-inch (9-cm) telescope, M3 is a fuzzball with a hint of granularity; in an 8-inch (20-cm) at 140×, it is extensively resolved.

---

41 **17 Comae Berenices** DOUBLE STAR IN COMA BERENICES
SAO 82330 12^h28.9^m +25°55′
NexStar Star 2968 Magnitudes 5.2, 6.6
Separation 145″
Position angle 250°

This very wide double, best seen at very low power (< 40×), is in the middle of the **Coma Berenices star cluster**, 10° in diameter. The cluster is prominent to the unaided eye or in the finder, but is much too large to fit in the telescopic field.

---

42 **Black Eye Galaxy** SPIRAL GALAXY IN COMA BERENICES
M64 12^h56.7^m +21°41′
NGC 4826 Magnitude 8.5
Size 10′ × 5′

The "black eye" in this bright galaxy is a dark dust lane just south of center. The galaxy is easy to see; the dust lane is something of a challenge. As with other galaxies, a dark country sky is required for a good view.

---

43 γ **Leonis**   DOUBLE STAR IN LEO

SAO 81298   10^{h}20.0^m +19°51′

HIP 50583   Magnitudes 2.4, 3.6

LX200 Star 102   Separation 4.3″

NexStar Star 2560   Position angle 124°

Gamma Leonis is also known as **Algeiba**. It is a fine bright double with stars exactly matched in color though not in brightness. Use 100× or more.

---

44   **M65**   SPIRAL GALAXY IN LEO

NGC 3623   11^{h}18.9^m +13°05′

                                                 Magnitude 8.8

                                                 Size 10′×3′

45   **M66**   SPIRAL GALAXY IN LEO

NGC 3627   11^{h}20.2^m +12°59′

                                                 Magnitude 9.0

                                                 Size 9′ × 4′

46   **NGC 3628**   SPIRAL GALAXY IN LEO

                                             11^{h}20.3^m +13°35′

                                                Magnitude 10.5

                                                 Size 15′ × 3.5′

These three galaxies (the **Leo Triplets**) fit in the same field at 40× or lower. M65 and M66 are among the most often observed galaxies; I have seen them in 8 × 40 binoculars from Kitt Peak and in a 7 × 50 monocular from Georgia. NGC 3628 is distinctly fainter than the others.

---

47   **M60**   ELLIPTICAL GALAXY IN VIRGO

NGC 4649   12^{h}43.7^m +11°33′

                                                 Magnitude 8.8

                                                 Size 7.4′×6.0′

The elliptical galaxy M60 is the brightest of a chain of galaxies extending to the northwest through M59, M58, M89, M87, M90, M88, M91, M86, M84, M99, M98, M100, M85, and numerous NGC galaxies. You may see as many as four or five galaxies, some bright and some fainter, in a single low-power field.

In particular, NGC 4647, 12th magnitude, is just northwest of M60; the pair reminds O'Meara of Mizar and Alcor.

A dark country sky is required.

---

48 3C 273   QUASAR IN VIRGO
        PGC 41121   12^{h}29.1^m +2°04′
                                                                     Magnitude 12.2–13.0

Challenging object. This is the only quasar visible with an 8-inch telescope, and even then, not an easy object; it looks like a faint bluish star of magnitude 12.5, close to a 13.5-magnitude star with which it seems to form a double (Figure 14.2).

Since it has no NGC number, with most telescopes you must find it by its coordinates (in full: 12^{h}29^{m}07^s +02°03′09″). A 10th-magnitude star is about 8′ to the southeast. A degree to the southeast is the fine double star h 146 (the next object in this list).

Under a suburban sky I needed medium power (140×) to see 3C 273 in an 8-inch (20-cm); lower powers showed field stars down to about magnitude 12, but not 3C 273 or its 13th-magnitude companion.

Like all quasars, 3C 273 has a strongly redshifted spectrum. This does not make it look reddish because the blue parts of the visible spectrum are filled in by light shifted from the ultraviolet range. If due to the expansion of the universe – as it appears to be – 3C 273's redshift implies a distance of 3000 million light years, probably the most distant object (and most ancient photons) you will ever see. Other quasars have even stronger redshifts, implying distances up to 12 000 million light years.

---

49 h 146   DOUBLE STAR IN VIRGO
        ADS 8582   12^{h}31.2^m +1°20′
        SAO 119447   Magnitudes 7.7, 8.7
        HIP 61091   Separation 49″
                                                                   Position angle 290°

This fine double star is of interest mainly as a guidepost to 3C 273 (the previous object in this list). It is too faint to be in built-in catalogues and must be found by coordinates (in full: 12^{h}31^{m}15^s +01°19′37″). The brighter star looks slightly warmer-colored, perhaps just because it is brighter.

---

50 γ **Virginis**   DOUBLE STAR IN VIRGO
        SAO 138917   12^{h}41.7^m − 1°27′
        HIP 61941   Magnitudes 2.7, 2.8
        LX200 Star 129   See orbit diagram
        NexStar Star 3016

Gamma Virginis is also called **Porrima**. This double star has a rapid 169-year orbit seen edge-on, so the separation is rapidly decreasing (Figure 14.3). In the

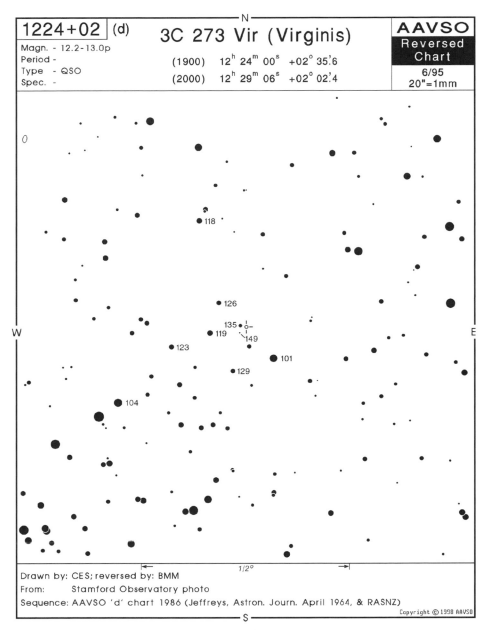

Figure 14.2. Magnitude comparison chart for 3C 273, showing telescopic view at medium power, flipped to match the view through a telescope with a diagonal. Magnitudes are shown with the decimal point omitted; thus 135 means 13.5.

1970s, it was an easy object, and Burnham compared it to "the twin head-lamps of some celestial auto." A well-collimated 8-inch (20-cm) telescope resolved it in 2001, but in 2005 probably will not be able to do so. Catch it while you can. I split it with a NexStar 5 in 2001 but will probably not do so again until at least 2008.

γ Virginis (WDS)

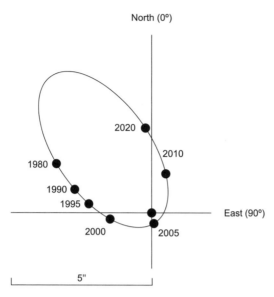

Figure 14.3. The orbit of γ Virginis (Porrima), flipped to match the view through a telescope with a diagonal.

Because of its high proper motion and relatively rapid orbit, this star is not cross-indexed properly in computerized databases; SIMBAD lists it as four objects, and *TheSky* version 5 also treats it inconsistently.

---

51 **Spindle Galaxy** ELLIPTICAL GALAXY IN SEXTANS
NGC 3115 10ʰ05.2ᵐ −7°43′
Caldwell 53 Magnitude 8.9
Size 8.3′×3.2′

This elongated elliptical galaxy, seen from the side, looks very narrow and has a relatively bright core. It is about 25 million light-years away.

"An easy telescope object, even for urban skies" (Ratledge), at least if your "urban skies" are not too bright; about equal in visibility to M82. Use medium power. Under really bad skies, high power may help.

---

52   **Sombrero Galaxy**   SPIRAL GALAXY IN VIRGO
         M104 $12^h40.0^m$ $-11°37'$
         NGC 4594   Magnitude 8.0
                                                                  Size 8.7′ × 3.5′

This relatively bright galaxy looks like a blurry image of Saturn; in larger telescopes and photographs, a dust lane makes it look like a Mexican hat. It is
65 million light years away, more than 30 times the distance of M31.

Because of its high surface brightness, M104 is a good choice for observing under less than ideal conditions. Part of the dust lane is visible in an 8-inch
(20-cm) at 140× under suburban sky conditions. Viewed with the 16-inch
(40-cm) Meade LX200 on Kitt Peak in Arizona, it is a spectacular sight.

---

53   **Σ 1604**   TRIPLE STAR IN CORVUS
         Sinnott 5 $12^h09.5^m$ $-11°51'$
         SAO 157111   Magnitudes 6.5, 9.1, 8.4
         HIP 59272   Separations 9.4″, 11.6″
                                                                 Position angles 89°, 46°

An unusual triple star forming a right triangle. The stars do not all form a system
in space. Best seen in a 5-inch (12.5-cm) or larger telescope.

Σ stands for F. G. W. Struve, who catalogued double stars about a century
ago.

---

54   **Stargate**   ASTERISM IN CORVUS
         SAO 157385 $12^h36.9^m$ $-12°04'$
         HIP 61486   Magnitude 6
         NexStar Star 8063

A fine sight in the ETX-90 at low power. Almost in the field with M104, this
asterism is a striking triangle within a triangle, giving an illusion of perspective.
According to Harrington (*The Deep Sky*), it was dubbed *Stargate* by John Wagoner,
presumably because it looks like the fictional space portal in the 1994 movie of
that name.

---

55 δ **Corvi**
SAO 157323
HIP 60965
LX200 Star 123
NexStar Star 2971

DOUBLE STAR IN CORVUS
12^{h}29.9^m −16°31′
Magnitudes 3.0, 8.5
Separation 24″
Position angle 216°

Delta Corvi, also called **Algorab**, is a double star variously described as "white and orange" or "yellowish and pale lilac." Both stars look white to me. Note the great difference in brightness.

---

56 **R Crateris**
GCVS 290001
SAO 156389
HIP 53809

SEMIREGULAR VARIABLE STAR IN CRATER
11^{h}00.5^m −18°18′
Magnitude 7.5–9.0

This star is just 11′ east of α Crateris (SAO 156375, NexStar Star 2691), which is much brighter.

The color of this semiregular variable is something of a puzzle. To me it looks topaz-colored or golden, not red. It was described as "scarlet" by John Herschel (1847) and as "ruby" by T. W. Webb, but its spectral class, M4, would lead one to expect only a mild reddish color, and Hipparcos found a color index of only 1.4. Has the color varied? Does interstellar gas give it an odd tinge? See *Sky & Telescope*, March, 1970, p. 204.

---

57 **Ghost of Jupiter**
NGC 3242
Caldwell 59

PLANETARY NEBULA IN HYDRA
10^{h}24.8^m −18°38′
Magnitude 8.6
Size 40″×35″

Prominent object, even in relatively small telescopes. This bright, oval-shaped planetary nebula looks like a ghost image of Jupiter, hence its name. Note its unusual bluish color.

---

58 **Ring-Tail Galaxy**
The Antennae
NGC 4038, 4039
Caldwell 60

COLLIDING GALAXIES IN CORVUS
12^{h}01.9^m −18°53′
Magnitude 10.3
Size 2.6′×1.8′

Challenging object – requires a good, clear sky. This strange object is now believed to be a pair of colliding galaxies; Bart J. Bok once described them as "two spirals having a fight of some kind." At least an 8-inch (20-cm) telescope is needed to see the unusual structure.

---

59 **3 Centauri**   DOUBLE STAR IN CENTAURUS
SAO 204917   13^{h}51.8^m −33°00′
HIP 67669   Magnitudes 4.5, 6.0
NexStar Star 3253   Separation 7.0″
                                          Position angle 108°

Most people don't realize the constellation Centaurus extends this far north. In ancient times it was visible from Europe; precession has shifted it southward. Unfortunately, its most famous star, α Centauri, is not visible from the continental United States.

This double comprises two white stars, matched in color though not in brightness. The brighter one is very slightly bluer.

---

60 **Eight-Burst Nebula**   PLANETARY NEBULA IN VELA
NGC 3132   10^{h}07.0^m −40°26′
Caldwell 74   Magnitude 8.2
                                          Size 1.4′ × 0.9′

Most planetary nebulae block the view of the star that illuminates them, but you will be able to see the 11th-magnitude central star in this one. Around the star is a nebula of moderate (not high) surface brightness; the whole structure looks like a star surrounded by fuzz. "Eight-burst" refers to its structure as shown on photographs.

---

61 **Centaurus A**   GALAXY IN CENTAURUS
NGC 5128   13^{h}25.5^m −43°01′
Caldwell 77   Magnitude 7
                                          Diameter 18′

This bright, large galaxy is so far south that in the United States only the southernmost parts of the country get a good view of it, though Rich Jakiel has observed it from New York State. Clear air and a dark sky are required.

This galaxy is a strong radio source. In photographs or a large telescope, it looks like a ball with a dark dust lane across it. It is classified as type S0, between spiral and elliptical.

---

62 ω Centauri   GLOBULAR CLUSTER IN CENTAURUS
         NGC 5139   13^{h}26.8^m −47°29′
         Caldwell 80   Magnitude 3.5
                                               Diameter 36′

Although it is the finest globular cluster in the sky, Omega Centauri is so far south that in the U.S. only the southern states can see it reasonably well, and even then only for a short time each day, when it is highest. It is due south of Spica and is visible when Spica is on the meridian.

   This is the largest and most massive globular cluster in our galaxy, ten times the size of ordinary globulars, and if it were farther away, we would not hesitate to classify it as a compact dwarf elliptical galaxy.

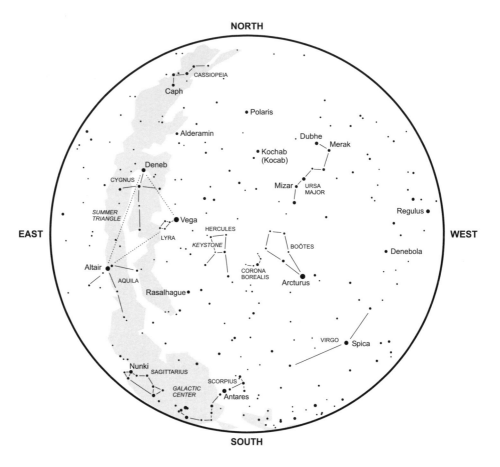

THE SKY AT SIDEREAL TIME 16:00

Mid-May	1 a.m. daylight saving time
Mid-June	11 p.m. daylight saving time
Mid-July	9 p.m. daylight saving time

As seen from latitude 40° north.

British and Canadian observers will see more stars above the north horizon;
southern U.S. observers will see more stars above the south horizon.

Based on a chart created with *Starry Night Pro* astronomy software
(http://www.starrynight.com), reproduced by permission.

Chapter 15
The May–June sky (R.A. 14^h–18^h)

63 **Cat's Eye Nebula** PLANETARY NEBULA IN DRACO
NGC 6543 17^{h}58.6^m +66°38′
Caldwell 6 Magnitude 8.3
Size 22″×16″

A surprisingly bright planetary nebula half the apparent size of Jupiter, oval or eye-shaped. The "cat's eye" appearance is evident only on large observatory photos.

    I find it clearly non-stellar even at 50×, but you may need to use higher power to be sure it is not a star.

---

64 **M92** GLOBULAR CLUSTER IN HERCULES
NGC 6341 17^{h}17.1^m +43°08′
Magnitude 6.5
Diameter 14′

As Burnham and O'Meara note, if M92 were not so close to M13 it would be considered a showpiece. As it is, not only does M13 overshadow it, but M92 is hard to find without computer aid because there are no bright stars nearby. Thus, many amateurs have never viewed it. Have a look – it's surprisingly bright and well worth seeing.

---

65 **Hercules Cluster** GLOBULAR CLUSTER IN HERCULES
M13 16^{h}41.7^m +36°27′
NGC 6205 Magnitude 5.3
Diameter 21′

Prominent object. Perhaps the finest globular cluster in the northern sky, M13 is easily resolved in a 5-inch (12.5-cm) or larger telescope. Like other globulars, it looks much more three-dimensional in the telescope than in photographs because the eye can see brightness distinctions that the camera does not record. Look for streams and other patterns in the stars.

---

66 **R Coronae Borealis**   VARIABLE STAR IN CORONA BOREALIS
GCVS 270001   15ʰ48.6ᵐ +28°10′
SAO 84015   Magnitude 5.7–14.8
HIP 77442
NexStar Star 3662

R Coronae Borealis is normally 6th magnitude but unpredictably fades to 12th magnitude or fainter for a period of days to weeks. Compare it to the 7.4-magnitude star 22′ to the north–northwest, in the same low-power field.

---

67 **Izar (Mirak)**   DOUBLE STAR IN BOÖTES
ε Boötis   14ʰ45.0ᵐ +27°04′
SAO 83500   Magnitudes 2.6, 4.8
HIP 72105   Separation 2.6″
LX200 Star 154   Position angle 341°
NexStar Star 3413

Epsilon Boötis has an abundance of names. In antiquity it was Izar or Mirak. Later, F. G. W. Struve named this close yellow-and-blue double star **Pulcherrima** (Latin for "most beautiful"). It requires steady air and high power.

---

68 **T Coronae Borealis**   RECURRENT NOVA IN CORONA BOREALIS
GCVS 270003   15ʰ59.5ᵐ +25°55′
SAO 84129   Magnitude 2.0–10.8
HIP 78322
LX200 Star 170

This star is normally magnitude 10.8 (fainter than the 9.3- and 9.8-magnitude stars 15′ to the west and northeast of it) but occasionally flares up to naked-eye visibility; there have been three outbursts since 1970, each of them very short. Take a look and see what it's doing tonight.

Although it has an SAO number, this star is too faint to be in most telescopes' built-in catalogues and must be found by its coordinates (in full: 15ʰ59ᵐ30ˢ +25°55′13″).

There is a fine 10th-magnitude double star within the same low-power field, 15′ to the southeast.

---

69 NGC 6210

PLANETARY NEBULA IN HERCULES
16^{h}44.5^m +23°49′
Magnitude 9
Diameter 20″

A splendidly bright planetary nebula, perhaps smaller than its listed size. Use medium to high power; below 100× it is easily mistaken for a star. There is a reddish 7th-magnitude star in the field.

---

70 ξ **Boötis**
SAO 101250
HIP 72659
LX200 Star 312
NexStar Star 3445

DOUBLE STAR IN BOÖTES
14^{h}51.4^m +19°06′
Magnitudes 4.5, 6.8
See orbit diagram

Xi Boötis is a beautiful double star with a color contrast, described as yellow and red-tinged, though both stars have nearly the same color index; a fine sight even at low power.

The orbit is relatively rapid (151 years) so that the position angle changes appreciably within one observer's lifetime (Figure 15.1). The stars were aligned north–south around 1950 and will be aligned east–west around 2035.

---

71 β **Serpentis**
SAO 101725
HIP 77233
NexStar Star 3653

DOUBLE STAR IN SERPENS CAPUT
15^{h}46.2^m +15°25′
Magnitudes 3.7, 10.0
Separation 31″
Position angle 264°

Beta Serpentis is a fine double, well seen at low power in 5-inch (12.5-cm) or larger telescopes, variously described as "both blue" or "blue and yellow."

---

72 **Rasalgethi**
α Herculis
SAO 102680
HIP 84345
LX200 Star 190
NexStar Star 3996

DOUBLE STAR IN HERCULES
17^{h}14.6^m +14°23′
Magnitudes 3.5, 5.4
Separation 4.8″
Position angle 105°

Rasalgethi, alias Alpha Herculis, is a fine close pair, yellow and blue.

ξ Boötis (WDS)

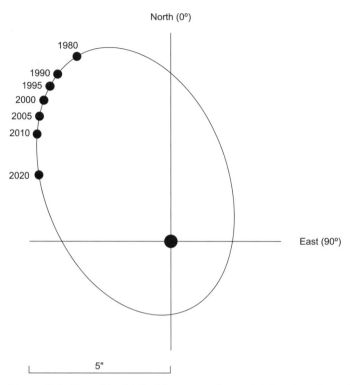

Figure 15.1. The orbit of ξ Boötis, reversed to match the view through a telescope with a diagonal.

Pluto (LX200 Star 909) is in this region through 2020. At magnitude 14, it is just within reach of an 8-inch (20-cm) telescope at a very dark site, and faint but plainly visible in a 16-inch (40-cm) against a dark sky.

---

73   δ **Serpentis**   DOUBLE STAR IN SERPENS CAPUT
          SAO 101623   15^{h}34.8^m +10°32′
          HIP 76276   Magnitudes 4.2, 5.2
          LX200 Star 317   Separation 4.1″
          NexStar Star 3600   Position angle 174°

Delta Serpentis is a fine close double comprising two white stars. Use medium to high power.

---

74   IC 4665

OPEN CLUSTER IN OPHIUCHUS
17^{h}46.3^m +5°43′
Magnitude 4
Diameter 40′

Prominent object. This neglected but bright open cluster is one of my personal favorites, mainly because it is so little known. Not only did Messier miss it, but so did all the observers whose work was consolidated to make the NGC. It more than fills the medium-power field and is a good object for small telescopes at low power.

---

75   Barnard's Star
HIP 87937

HIGH-PROPER-MOTION STAR IN OPHIUCHUS
17^{h}57.8^m +4°42′
Magnitude 9.5

Not in the LX200's built-in catalogue, this star must be found by right ascension and declination followed by careful comparison to the chart on p. 96.

At 6 light-years, Barnard's Star is the nearest star other than the Sun and the α Centauri system. It is also the star with the largest known proper motion, about 1° in 350 years; its motion is plotted in *Sky Atlas 2000.0*, second edition. Movement relative to the field stars is noticeable over periods as short as 20 years. Make a careful sketch now and return to this star in your old age.

---

76   U Scorpii
GCVS 730004

RECURRENT NOVA IN SCORPIUS
16^{h}22.5^m −17°53′
Magnitude 8.7–19.3

You should not see a star here, but have a look just in case. U Scorpii is a recurrent nova that flared up in 1979 and 1998. Compare it to several 10th- and 11th-magnitude stars in the field.

This star is not in the Meade LX200's built-in GCVS catalogue. Its precise position is 16^{h}22^{m}31^s −17°52′42″.

---

77 ν Scorpii
SAO 159763
HIP 79374
NexStar Star 3766

QUADRUPLE STAR IN SCORPIUS
16^{h}12.0^m −19°28′
Magnitudes 4.3, 5.3, 6.6, 7.2
Separations: AB 1.3″, AC 41″, CD 2.4″
Position angles: AB 2°, AC 336°, CD 54°

Nu Scorpii is an easy 41″ double, or a quadruple star more challenging than ϵ Lyrae, depending on your point of view. It is resolvable into four stars with a well-collimated 5-inch (12.5-cm) or larger telescope in steady air. The close pairs have widened appreciably since they were discovered in the nineteenth century.

---

78   **Antares**   VARIABLE DOUBLE STAR IN SCORPIUS
         α Scorpii   16^h29.4^m −26°26′
         SAO 184415   Magnitudes 0.9–1.8, 5.4
         HIP 80763   Separation 2.8″
         LX200 Star 177   Position angle 277°
         NexStar Star 3840

Challenging object. Antares is both a semiregular red giant variable and a binary. Its companion is cooler and is often described as green in color, an illusion caused by contrast. High power and steady air are required to see the companion.

---

79   **M4**   GLOBULAR CLUSTER IN SCORPIUS
         NGC 6121   16^h23.6^m −26°31′
                                                                        Magnitude 5.4
                                                                        Diameter 35′

This unusually nearby globular cluster, only a quarter as far away as M13, is easily resolved into stars even in small telescopes. Partly because it is a relatively loose cluster, and partly because there is dark nebulosity in front of it, M4 appears to be criscrossed by vague, dark streamers.

The small 9th-magnitude globular cluster NGC 6144 is just north of a point halfway between M4 and Antares.

---

80   **Ptolemy's Cluster**   OPEN CLUSTER IN SCORPIUS
         M7   17^h53.8^m −34°47′
                                                                        Magnitude 2.8
                                                                        Diameter 1.5°

Prominent object. This bright open cluster was noted by Claudius Ptolemy over 1800 years ago and is a fine sight in small telescopes at low power. There is dark nebulosity in the region; look for elongated starless regions just outside the cluster.

Other Messier objects in this area include M6 (open cluster) and M19 and M62 (globulars).

---

81   **Caldwell 76**   OPEN CLUSTER IN SCORPIUS
        NGC 6231   16^{h}54.0^m −41°48′
                                                    Magnitude 2.6
                                                    Diameter 15′

Prominent object, neglected only because of its southerly location. "A fabulous sparkling open cluster" (Ratledge), roughly triangular in shape. Containing large numbers of bright blue-white stars, it physically resembles the Pleiades.

---

82   **μ Lupi**   QUADRUPLE STAR IN LUPUS
        SAO 225638   15^{h}18.5^m −47°53′
        HIP 74911   Magnitudes 4.9, 5.0, 7.6, 7.0
        NexStar Star 3538   Separations: AB 1.0″, AC 24″, AD 241″
                                   Position angles: AB 128°, AC 129°, AD 248°

Mu Lupi is an intriguing quadruple star with companions near and far. The most distant companion (component D, catalogued as SAO 225627) is visible in binoculars. All four stars apparently form a physical system.

At mid-northern latitudes μ Lupi is above the horizon only briefly; it is due south of β Librae (SAO 140430).

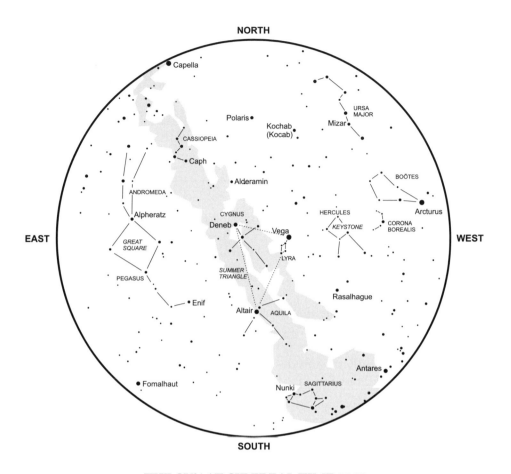

THE SKY AT SIDEREAL TIME 20:00

Mid-July	1 a.m. daylight saving time
Mid-August	11 p.m. daylight saving time
Mid-September	9 p.m. daylight saving time

As seen from latitude 40° north.

British and Canadian observers will see more stars above the north horizon;
southern U.S. observers will see more stars above the south horizon.

Based on a chart created with *Starry Night Pro* astronomy software
(http://www.starrynight.com), reproduced by permission.

Chapter 16
The July–August sky (R.A. 18^h–22^h)

This chapter has the lion's share of deep-sky objects because it covers the richest part of the sky. On summer evenings we are looking toward the center of our galaxy (in Sagittarius) and one of its spiral arms (in the direction of Cygnus). Fortunately, in the summer most of us have more time to observe, and the weather is warm if hazy.

---

83 β **Cephei**   DOUBLE STAR IN CEPHEUS
SAO 10057   21^{h}28.7^m +70°34′
HIP 106032   Magnitudes 3.2, 8.6
LX200 Star 236   Separation 13.3″
NexStar Star 5055   Position angle 250°

Prominent object. Beta Cephei is a fine double, white and bluish-white. The primary is a very low-amplitude variable. This is a good star on which to practice measuring double stars because the separation and position angle have been unchanged at least since 1830.

---

84   **T Cephei**   LONG-PERIOD VARIABLE STAR IN CEPHEUS
GCVS 200003   21^{h}09.5^m +68°29′
SAO 19229   Magnitude 5.2–11.3
HIP 104451   Period 388 days
NexStar Star 4986

A very red Mira-type variable almost in the field with NGC 7023 (Caldwell 4, a faint reflection nebula). The period is a year plus almost a month. Maxima of T Cephei are predicted for September 2002, October 2003, November 2004, November (again) 2005, December 2006, and so on.

Compare to stars of magnitude 6.9 and 8.1 in the same low-power field, about 15′ to the southwest and northwest respectively.

---

85 Σ **2398**   DOUBLE RED DWARF IN DRACO
SAO 31128   18ʰ42.8ᵐ +59°38′
HIP 91768   Magnitudes 9.1, 10.0
                     Separation 13″
                     Position angle 173°

This moderately faint double star consists of two red dwarfs only 11 light-years away. There are several wider doubles in the field.

Red dwarfs are the most common stars, but we don't see very many because they are not very luminous. The brighter of the two stars is thought to be a flare star, which means it may sometimes appear substantially brighter than its companion.

Not listed in the built-in catalogues of most telescopes, Σ 2398 must be found by its coordinates (in full: 18ʰ42ᵐ46ˢ +59°37′52). Be sure to sync on a nearby star first.

Σ stands for F. G. W. Struve, who catalogued double stars about a century ago. This star is also known as Aitken Double Star (ADS) 11632 and as WDS 18432 + 5933. The latter designation disagrees with the Hipparcos position and was changed to WDS 18428 + 5938 in October 2001.

---

86 *o* Draconis   DOUBLE STAR IN DRACO
SAO 31219   18ʰ51.2ᵐ +59°23′
HIP 92512   Magnitudes 4.8, 8.3
NexStar Star 4412   Separation 37″
                     Position angle 319°

Omicron Draconis is not a real double star, but an optical pair, described as orange and blue.

---

87 **Garnet Star**   VARIABLE STAR IN CEPHEUS
μ Cephei   21ʰ43.5ᵐ +58°47′
SAO 33693   Magnitude 3.4–5.1
HIP 107259
NexStar Star 5107

Mu Cephei is a semiregular variable named the Garnet Star by Sir William Herschel because of its unusual golden or reddish color (spectral type M, color index 2.7). It is actually a triple; there are 12th-magnitude companions at

19″ and 40″, both of which should be visible with an 8-inch (20-cm) under a good dark sky.

---

88   IC 1396

OPEN CLUSTER IN CEPHEUS
21ʰ39.1ᵐ +57°30′
Magnitude 3.5

A loose cluster of stars with a triple star on one side and a double on the other, filling the field at 50×. Large telescopes at dark sites will also show nebulosity. The cluster surrounds the triple star SAO 33626 (NexStar Star 5085, mags. 5.7, 8.1, 8.0, seps. 12″, 20″, p.a. 120°, 324°).

---

89   **Blinking Planetary**
NGC 6826
Caldwell 15

PLANETARY NEBULA IN CYGNUS
19ʰ44.8ᵐ +50°31′
Magnitude 9.8

The central star in this planetary nebula is much brighter than the surrounding nebulosity. The star and the nebula compete for your attention; the "blinking" effect occurs because the star pops in and out of view as you switch from direct to averted vision, or so the folklore says.

I don't see this effect myself; to my eyes, the star and the nebula do compete for attention, but neither one blinks out when attention is directed at the other. This may be because I am too accustomed to using averted vision; beginners see the illusion more easily.

The bright double star **16 Cygni** (SAO 31898, NexStar Star 9426, mags. 6.0, 6.2, sep. 40″) is just out of the field to the west. Several other doubles are nearby.

---

90   **Webb's Horseshoe**

ASTERISM IN CYGNUS
21ʰ08.0ᵐ +47°15′
Magnitude 7
Diameter 15′

This horseshoe-shaped asterism caught the eye of T. W. Webb in the nineteenth century, who called it a "curious horseshoe, and magnificent Galaxy [i.e., Milky Way] field." It consists of "an arc of about 14 suns ranging from about 7th to about 11th magnitude and spanning $\frac{1}{4}$°" (*Sky & Telescope*, May, 1998, p. 104).

Not catalogued as a cluster, this asterism must be found by its coordinates or by going to SAO 50456 (NexStar Star 4979) and slewing half a degree to the south. Also almost in the field is the bright, wide double star SAO 50536 (NexStar Star 9822, mags. 6.6 and 7.5, sep. 136″, p.a. 189°).

Arcs of stars are moderately common – the constellations Corona Borealis and Corona Australis are the most conspicuous examples – and it is an open question whether they have a physical basis, perhaps due to star formation along a shock wave traveling through interstellar gas.

---

91 **Patriotic Triple**   TRIPLE STAR IN CYGNUS
o^1 Cygni   20^{h}13.6^m +46°44′
31 Cygni   Magnitudes 3.9, 7.0, 4.8
SAO 49337   Separations 105″, 330″
HIP 99675   Position angles 174°, 324°
NexStar Star 4774

Prominent object. "A 3.8-magnitude reddish orange star flanked by a 6.7-magnitude clear blue star and a 4.8-magnitude white one, all set against a rich Milky Way field" (Mullaney). Use low power. This red, white, and blue asterism centered on omicron-1 Cygni does not form a true system in space.

---

92 **SS Cygni**   RECURRENT NOVA IN CYGNUS
GCVS 310019   21^{h}43.0^m +43°37′
                                                                  Magnitude 7.7–12

An extreme type of recurrent nova, this star is bright about 20% of the time. For a chart of comparison stars, see p. 141. Find SS Cygni by coordinates (in full: 21^{h}42^{m}43^s +43°35′10″) or by going to 75 Cygni (the next object in this list) and slewing half a degree to the northeast.

SS Cygni consists of a spectroscopic binary star with an unusually short orbital period, just 6 days. The two stars are orbiting almost in contact, and the smaller star refuels itself from the larger one, temporarily flaring up with greater light as it does so.

---

93 **75 Cygni**   TRIPLE STAR IN CYGNUS
SAO 51167   21^{h}40.2^m +43°16′
HIP 106999   Magnitudes 5.1, 9.8, 10.2
NexStar Star 5089   Separations 2.8″, 63″
                                                         Position angles 326°, 254°

A star with both nearby and distant companions, almost in the field with SS Cygni (previous object).

---

94   **NGC 7027**

PLANETARY NEBULA IN CYGNUS
21^{h}07.1^m +42°14′
Magnitude 10
Size 10″ × 5″

A small, bright, elongated planetary nebula with a faint outer shell.

---

95   **χ Cygni**
SAO 68943
HIP 97629
NexStar Star 4668

LONG-PERIOD VARIABLE STAR IN CYGNUS
19^{h}50.6^m +40°36′
Magnitude 5.2–13.4
Period 408 days

Chi Cygni is a prominent long-period Mira-type variable, visible to the naked eye during part of its cycle, but sometimes beyond the reach of even an 8-inch (20-cm) telescope when at minimum. Maxima are expected in March 2003 (hard to observe because of the Sun), May 2004 (more favorable), June 2005 (convenient), and July 2006 (very convenient to observe).

---

96   **The Double Double**
$\epsilon^{1,2}$ Lyrae
SAO 67309, 67315
HIP 91919, 91926
LX200 Star 334, 335
NexStar Star 4374, 4376

QUADRUPLE STAR IN LYRA
18^{h}44.3^m +39°40′
Magnitudes 5.0, 6.1, 5.2, 5.5
Separations: AB 2.7″, AC 210″, CD 2.5″
Position angles: AB 348°, AC 174°, CD 159°

Prominent object. Epsilon Lyrae consists of two stars 210″ apart, each of which is a close double; all four form a single system in space. These stars are a good test of telescope optics. In steady air, a well-collimated 3.5-inch (9-cm) or larger telescope will show each pair clearly, with a distinct gap between the adjacent diffraction disks.

---

97   **61 Cygni**
SAO 70919
HIP 104214
LX200 Star 346
NexStar Star 4980

DOUBLE STAR IN CYGNUS
21^{h}06.9^m +38°45′
Magnitudes 5.4, 6.1
Separation 31″
Position angle 150°

This handsome double star was the first star whose distance was measured by parallax (by Bessel in 1838). At one time it was known as **Piazzi's Flying Star** because of its large proper motion (which has also created cross-indexing problems in some computerized star catalogues). Both components are yellowish, with a color index of about 1.

---

98   **Sinnott 10**   TRIPLE STAR IN CYGNUS
                                                         21ʰ35.1ᵐ +38°07′
                                                  Magnitudes 10.4, 10.6, 10.8
                                                        Separations 19″, 19″
                                                    Position angles 313°, 9°

The most perfect equilateral triple star in the heavens, according to Roger W. Sinnott, who discovered it by doing a computer search of the Hipparcos and Tycho catalogues; not previously noted as a multiple star. Find it by right ascension and declination, or by going to 72 Cygni (SAO 71480, NexStar Star 5071, mag. 5) and slewing half a degree to the south.

   The triple star looks like a nebulous patch at 50×; higher power reveals its true nature.

---

99   **ζ Lyrae**   DOUBLE STAR IN LYRA
         SAO 67321   18ʰ44.8ᵐ +37°36′
         HIP 91971   Magnitudes 4.3, 5.6
         NexStar Star 4377   Separation 41″
                                                        Position angle 154°

Prominent object. Zeta Lyrae is a fine double star at low power, due south of ε Lyrae. It is a good star on which to practice your measuring technique because the position angle and separation are essentially constant.

---

100   **T Lyrae**   CARBON STAR IN LYRA
          GCVS 520003   18ʰ32.3ᵐ +37°00′
          SAO 67087   Magnitude 7.8–9.3
          HIP 90883

An extremely red star, with color index 5.1 (Hipparcos) to 5.5 (Skiff). Compare it to V Aquilae (p. 216) and the Garnet Star (p. 206).

---

101   **Cygnus X-1**   STAR AND BLACK HOLE IN CYGNUS
          V1357 Cygni   19ʰ58.4ᵐ +35ʰ12ᵐ
          GCVS 311357   Magnitude 8.8

This white 9th-magnitude star, a low-amplitude variable, has an optically invisible companion that emits X-rays and is thought to involve a black hole.

Cygnus X-1 looks like a wide double star because there is a 10th-magnitude foreground star 55″ to the north.

To find Cygnus X-1, use the coordinates or go to η Cygni (SAO 69116, NexStar Star 4704) and slew half a degree to the west, until you see the pair of stars close together, at the apex of an asterism shaped rather like the constellation Triangulum.

You can't see the black hole, of course; it doesn't emit light but is detected indirectly because of radio emissions from matter falling into it. Contrary to popular impressions, a black hole does not have unlimited power to consume matter; its mass and gravitational pull are finite, so unless matter gets close enough to fall in, a black hole can lead an untroubled life in a stable orbit indefinitely.

---

102   Σ **2470**   DOUBLE STAR IN LYRA
SAO 67870   19ʰ08.8ᵐ +34°46′
HIP 94043   Magnitudes 7.0, 8.4
NexStar Star 9274   Separation 14″
                                                 Position angle 270°

103   Σ **2474**   DOUBLE STAR IN LYRA
SAO 67875   19ʰ09.1ᵐ +34°36′
HIP 94076   Magnitudes 6.8, 7.9
                                                      Separation 16″
                                                 Position angle 264°

This is Lyra's *other* "double double," a striking and surprising sight. These are two very similar double stars – matched in brightness, separation, and position angle – less than a quarter of a degree apart. They form a nearly equilateral triangle with a single star. Unlike ε Lyrae, these two doubles do not form a system in space; the southern pair is much closer to us than the northern pair, which means that it is, in real life, also smaller and fainter.

Σ stands for F. G. W. Struve, who catalogued double stars about a century ago.

---

104   β **Lyrae**   VARIABLE DOUBLE STAR IN LYRA
SAO 67452   18ʰ50.1ᵐ +33°22′
HIP 92420   Magnitudes 3.3–4.3, 6.7
NexStar Star 4408   Separation 46″
                                                 Position angle 150°

Prominent object. Beta Lyrae is a fine low-power double star suitable for small telescopes. The primary star is variable, with a period of 13 days. Two

10th-magnitude stars on the opposite side of the primary are probably also part of the system, making it a quadruple.

---

105 Ring Nebula
M57
NGC 6720

PLANETARY NEBULA IN LYRA
18^{h}53.6^m +33°02′
Magnitude 8.8
Diameter 76″

Prominent object at 30× or higher. This large, bright planetary nebula is not completely dark in the center. It is genuinely donut-shaped; M27 is the same type of object seen from the side. Large telescopes reveal a 13th-magnitude star at the center.

A chart of stars near M57, labeled with magnitudes, appeared in *Sky & Telescope,* September, 2001, p. 102. It is useful for gauging the limiting magnitude of your telescope or your photographs.

---

106 Albireo
β Cygni
SAO 87301
HIP 95947
LX200 Star 223
NexStar Star 4597

DOUBLE STAR IN CYGNUS
19^{h}30.7^m +27°58′
Magnitudes 3.3, 4.7
Separation 34″
Position angle 54°

Prominent object. Albireo (Beta Cygni) is a magnificent yellow-and-blue double star, visible as double even in 8× binoculars firmly mounted on a tripod. The primary and secondary are spectral types K3 and B8, a star entering the red giant phase and its slower-evolving, blue-white companion.

---

107 16 Vulpeculae
SAO 88098
HIP 98636
NexStar Star 4739

DOUBLE STAR IN VULPECULA
20^{h}02.0^m +24°56′
Magnitudes 5.8, 6.2
Separation 0.8″
Position angle 122°

Challenging object. This double star is a good test for an 8-inch (20-cm) telescope in steady air. At 500×, I saw it as two diffraction disks with dark sky clearly visible between them. It is at the Dawes limit for a 6-inch (15-cm).

---

108 **Dumbbell Nebula** PLANETARY NEBULA IN VULPECULA
M27 19ʰ59.6ᵐ +22°43′
NGC 6853 Magnitude 7.3
Size 8′ × 6′

Prominent object. This dumbbell-shaped nebula is believed to be a ring like M57, but viewed from the side. It is one of the finest planetary nebulae in the sky, as well as one of the largest (about $\frac{1}{8}°$ across) and one of the few whose structure is clearly visible in modest-sized telescopes. Surface brightness is relatively high; I can see it easily, as a small glowing patch, in 8 × 40 binoculars.

---

109 **Coathanger** OPEN CLUSTER OR ASTERISM IN VULPECULA
Brocchi's Cluster 19ʰ25.4ᵐ +20°11′
Collinder 399 Magnitude 3.6
SAO 104831 Diameter 1.5°
NexStar Star 4580

This prominent loose star cluster looks like a coathanger, i.e., a horizontal line with a hook attached to its middle. It is too large for the field of the telescope but is well seen in the 8 × 50 finder. SAO 104831 (5 Vulpeculae) is one of the stars in it.

The small 8th-magnitude open cluster NGC 6802 is just east of the horizontal part of the coathanger.

---

110 **M71** GLOBULAR CLUSTER IN SAGITTA
NGC 6838 19ʰ53.8ᵐ +18°47′
Magnitude 8.0
Diameter 7′

This cluster's status was formerly in doubt; some classified it as an open cluster, since it is easily resolved into stars. It is now known to be a globular, though an unusually nearby one.

---

111 **γ Delphini** DOUBLE STAR IN DELPHINUS
SAO 106475 20ʰ46.7ᵐ +16°07′
HIP 102531 Magnitudes 4.4, 5.0
LX200 Star 342 Separation 9.2″
NexStar Star 4898 Position angle 266°

Gamma Delphini is a fine double star, described by observers as gold and blue even though the actual difference of color index is small. I see both stars as yellowish. How do they look to you?

---

112   Pegasus Cluster   GLOBULAR CLUSTER IN PEGASUS
M15   21^{h}30.0^m +12°10′
NGC 7078   Magnitude 6.0
                                               Diameter 18′

Prominent object. This globular cluster is unusually bright and is a fine sight even in moonlight or under hazy skies. An 8th-magnitude star is only 6′ from the center.

Large telescopes and/or dark skies are needed to resolve M15. O'Meara reports that the stars appear to be arranged in streams.

---

113 π Aquilae   DOUBLE STAR IN AQUILA
SAO 105282   19^{h}48.7^m +11°49′
HIP 97473   Magnitudes 6.3, 6.8
NexStar Star 4663   Separation 1.4″
                                               Position angle 105°

Pi Aquilae is a good test for a 5- or 6-inch (12- or 15-cm) telescope in steady air. With an 8-inch (20-cm) at 250× I saw it very cleanly split. It is at the Dawes limit for a 3.5-inch (9-cm) telescope, such as a Meade ETX-90 or Questar, which should show it as two disks just touching, without a dark space in between.

---

114   Sinnott 7   TRIPLE STAR IN AQUILA
                                               18^{h}55.4^m +11°18′
                                               Magnitudes 10.7, 10.7, 11.2
                                               Separations 27″, 31″
                                               Position angles 205°, 266°

A triple star forming an equilateral triangle, number 7 in a list of nearly equilateral triples discovered by Roger W. Sinnott; not previously noted as a multiple star.

This object is not in built-in catalogues; find it by its coordinates (in full: 18^{h}55^{m}23^s +11°17′39″) or by going to SAO 104239 (mag. 6.5) and slewing half a degree to the north–northeast.

Note that the three components of Sinnott 7 are relatively faint. The trio requires more than 50× for clear recognition. There is, however, a brighter, less symmetrical triangle 17′ to the northwest, within the same low-power field.

---

115   NGC 6572   PLANETARY NEBULA IN OPHIUCHUS
                                                  18ʰ12.1ᵐ +6°51′
                                                      Magnitude 9.0
                                                       Diameter 8″

An unusually bright planetary nebula, easily mistaken for a star except for its
bluish-green color. Use medium to high power; at 80× it is almost indistinguishable from a star, but at 140× it is a fine object, brighter than anything else in the
field.

---

116   NGC 6633   OPEN CLUSTER IN OPHIUCHUS
                                                  18ʰ27.2ᵐ +6°31′
                                                      Magnitude 4.6
                                                      Diameter 30′

This bright open cluster fills the field at 140×.

---

117   ϵ **Equulei**   TRIPLE STAR IN EQUULEUS
          SAO 126428   20ʰ59.1ᵐ +4°18′
          HIP 103569   Magnitudes 6.0, 6.3, 7.0
          LX200 Star 344   Separations (1999): 0.8″, 11.0″
          NexStar Star 4950   Position angles (1999): 287°, 66°

At first sight, Epsilon Equulei is a double star, but the brighter component is itself
a very close double in a rapid orbit seen almost edge-on, so that the separation
is decreasing but the position angle is relatively constant (Figure 16.1).

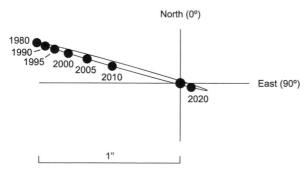

Figure 16.1. The orbit of ϵ Equulei A and B, flipped to match the view through a telescope
with a diagonal.

In 2000, the closer pair was just split by a well-collimated 8-inch (20-cm) telescope in very steady air. By 2010 it will be a challenge for a 16-inch (40-cm), and in 2018 it will probably not be resolvable by any Earth-based telescope.

---

118 **M2** GLOBULAR CLUSTER IN AQUARIUS
NGC 7089 21^{h}33.5^m − 0°49′
Magnitude 6.3
Diameter 16′

Another fine globular cluster, almost due south of M15. Under good conditions, M2 is easily resolved into stars with an 8-inch (20-cm) telescope. It is a fine sight in much smaller telescopes.

---

119 **Alvan Clark 11** DOUBLE STAR IN SERPENS CAUDA
SAO 142294 18^{h}25.0^m − 1°35′
HIP 90253 Magnitudes 6.7, 7.2
NexStar Star 4295 Separation 0.9″
Position angle 357°

Discovered by the famous Boston telescope maker Alvan Clark while testing his instruments, this close double star is a stiff test for a 6-inch (15-cm) or larger telescope. Use high power.

---

120 **V Aquilae** SEMIREGULAR VARIABLE STAR IN AQUILA
GCVS 050005 19^{h}04.4^m − 5°41′
SAO 142985 Magnitude 6.6–8.4
HIP 93666
NexStar Star 9254

V Aquilae is one of the reddest stars in the sky. It is the second reddest star in *Sky Catalogue 2000.0*, which gives its color index as 4.19, but Skiff gives it as 5.5.

---

121 **Wild Duck Cluster** OPEN CLUSTER IN SCUTUM
M11 18^{h}51.1^m − 6°16′
NGC 6705 Magnitude 5.3
Diameter 13′

Prominent object. This dense open cluster resembles a comet or a flock of wild ducks, hence the name. "A fluid fan of starlight streaming from an 8th-magnitude saffron star near the fan's apex" (O'Meara).

---

122 **Saturn Nebula**   PLANETARY NEBULA IN AQUARIUS
NGC 7009   21ʰ04.2ᵐ − 11°22′
Caldwell 55   Magnitude 8.3
                                                        Size 28″ × 23″

Prominent object at medium power. This famous bright planetary nebula looks somewhat like a ghost of Saturn. Larger telescopes show the "handles" at the ends.

---

123 **Eagle Nebula**   OPEN CLUSTER AND NEBULA IN SERPENS CAUDA
M16   18ʰ18.8ᵐ − 13°48′
NGC 6611   Magnitude 6
                                                       Size 120′ × 25′

This object was made famous by a Hubble Space Telescope photograph of three pillars of dark gas against a lighter background. You won't see *that* in a moderate-sized telescope; indeed, Messier classified M16 as a cluster without nebulosity. The luminous gas is relatively faint and requires a dark sky background.

The convoluted shape of this object has suggested a number of other names; it is also known as the **Ghost Nebula** or **Star Queen Nebula**.

---

124 **Little Gem**   PLANETARY NEBULA IN SAGITTARIUS
NGC 6818   19ʰ44.0ᵐ − 14°09′
                                                       Magnitude 10
                                                      Diameter 20″

This planetary nebula is a fine sight in the 8-inch (20-cm) at 140× even in full moonlight.

Neptune (LX200 Star 908) is in this region through 2020. At magnitude 8 and angular diameter 2″, it is easily mistaken for a star at low power, but look for its very un-starlike bluish-green color.

On December 21, 2009, Neptune and Jupiter will pass within a degree of each other. The same thing happened in 1613; at the time Galileo saw and recorded Neptune, though he thought it was a star.

---

125 **Omega Nebula** OPEN CLUSTER AND EMISSION NEBULA IN SAGITTARIUS
M17 18^{h}21.1^m $-$16°11′
NGC 6618 Magnitude 6
Diameter 25′

Prominent object. Unlike M16, this star cluster includes some quite bright nebulosity. It is also called the **Horseshoe Nebula** or **Swan Nebula** because of its shape, although to me it looks more like the digit 2 with an elongated base. The surface brightness is higher than M8 although the nebula is more compact. M17 is a good choice when you want to view an emission nebula under adverse conditions.

Numerous other Messier objects are in the vicinity, including M18, M23, M24, and M25.

---

126 **h 2866** TRIPLE STAR IN SAGITTARIUS
Sinnott 8 19^{h}23.4^m $-$18°00′
SAO 162554 Magnitudes 8.7, 8.7, 9.6
Separations 23″, 35″
Position angles 53°, 104°

A triple star forming a compact right triangle, discovered by Sir John Herschel; a fine sight at 50×. It is not in built-in catalogues. To find it, use its coordinates (in full: 19^{h}23^{m}6^s $-$17°59′58″) or go to ρ^1 Sagittarii (SAO 162512, NexStar Star 4556) and slew half a degree to the east.

---

127 **Trifid Nebula** OPEN CLUSTER AND NEBULA IN SAGITTARIUS
M20 18^{h}02.5^m $-$23°02′
NGC 6514 Magnitude 6.3
Diameter 20′

"Trifid" means "split into three," describing the shape of the gas cloud that accompanies this star cluster, apparently part of the same nebula as M8 to the south. It has also been called the **Clover**.

On photographs M20 is partly reddish, where the nebula is fluorescent, and partly blue-white, where it merely reflects starlight; it is crisscrossed by dark lanes. The open cluster M21 is 45′ to the northeast.

---

128 **Sagittarius Globular Cluster**   GLOBULAR CLUSTER IN SAGITTARIUS
M22   18^{h}36.4^m −23°54′
NGC 6656   Magnitude 5.2
Diameter 33′

Prominent object. One of the finest globulars in the sky, this cluster is bright and easy to resolve; in the 8-inch (20-cm) at 100× it is a shimmering ball of stars. In the North American sky the shimmer is largely caused by atmospheric turbulence, since M22 is never high.

---

129 **Lagoon Nebula**   OPEN CLUSTER AND NEBULA IN SAGITTARIUS
M8   18^{h}03.8^m −24°23′
NGC 6523   Magnitude 3
Size 90′ × 40′

Prominent object. After η Carinae (not visible from the United States) and M42, M8 is the third finest emission nebula in the sky. This is a region of active star formation, and indeed, this nebula is wrapped around a fine star cluster. The "lagoon" is the dark lane that cuts across the middle of the nebula.

---

130 **Loch im Himmel**   DARK NEBULA IN SAGITTARIUS
Near NGC 6520   18^{h}03.4^m −27°54′

When viewing this field, Sir William Herschel is reported to have exclaimed, "Hier ist wahrhaftig ein Loch im Himmel!" ("Here is truly a hole in the sky"). He thought the starless region to the south and east of the small open cluster NGC 6520 (mag. 8, diameter 5′) was a gap in our galaxy through which one could peer into intergalactic space. Today we know it as a dark nebula, Barnard 86.

---

131 **M54**   GLOBULAR CLUSTER IN SAGITTARIUS DWARF GALAXY
NGC 6715   18^{h}55.1^m − 30°29′
Magnitude 7.6
Diameter 9.1′

This unimportant-looking globular cluster is not part of our own galaxy. Instead, it belongs to the Sagittarius dwarf galaxy, a very faint, sparse galaxy discovered in 1994 which "our Milky Way Galaxy is now cannibalizing" (O'Meara).

According to Mallas and Kreimer, M54 is partly resolved even in a 4-inch (10-cm) telescope.

---

132 **NGC 6723**

The interesting thing about this globular cluster is the field around it. Rich Jakiel reports that even small telescopes show several dark nebulae (seen as starless regions) and, under good conditions, the reflection nebula NGC 6726/6729, which surrounds the 7th-magnitude star SAO 210828 half a degree to the southeast of the globular.

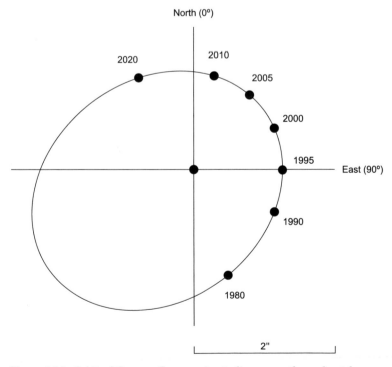

Figure 16.2. Orbit of Gamma Coronae Australis as seen through a telescope with a diagonal.

133 γ **Coronae Australis**   DOUBLE STAR IN CORONA AUSTRALIS
SAO 210928   19^{h}06.4^m − 37°04′
HIP 93825   Magnitudes 4.2, 4.3
LX200 Star 337   Separation ≈1.2″
NexStar Star 4493   See orbit diagram

Gamma Coronae Australis is a binary star in a 122-year orbit seen face-on, so that the position angle changes but the separation remains relatively constant (Figure 16.2).

This is a close double and requires steady air and high power. Note that in North America steady air may be elusive because this star does not rise very high in the sky.

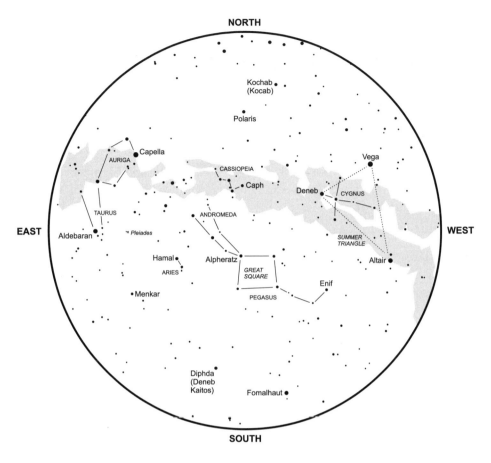

THE SKY AT SIDEREAL TIME 0:00

Mid-September	1 a.m. daylight saving time
Mid-October	11 p.m. daylight saving time
Mid-November	8 p.m. local time

As seen from latitude 40° north.

British and Canadian observers will see more stars above the north horizon;
southern U.S. observers will see more stars above the south horizon.

Based on a chart created with *Starry Night Pro* astronomy software
(http://www.starrynight.com), reproduced by permission.

Chapter 17
The September–October sky
(R.A. 22^h–2^h)

134   **Caldwell 10**   OPEN CLUSTER IN CASSIOPEIA
        NGC 663   1^{h}46.0^m +61°15′
                                                      Magnitude 7.1
                                                      Diameter 16′

Ratledge considers this cluster a finer sight than M103 nearby. In the 8-inch (20-cm) at 140×, it is a quite presentable open cluster containing a double star.

---

135   **M103**   OPEN CLUSTER IN CASSIOPEIA
        NGC 581   1^{h}33.4^m +60°40′
                                                      Magnitude 7.4
                                                      Diameter 6′

This rather faint cluster is a roughly equilateral triangle filled with stars.

---

136   **NGC 7510**   OPEN CLUSTER IN CEPHEUS
                                                      23^{h}11.5^m +60°34′
                                                      Magnitude 7.9
                                                      Diameter 3′

An unusual, compact, wedge-shaped open cluster, best seen at medium power.

---

137 **WZ Cassiopeiae** VARIABLE DOUBLE STAR IN CASSIOPEIA
GCVS 180048 0^{h}01.3^m +60°21′
SAO 21002 Magnitudes 6–8, 8.3
Separation 58″
Position angle 89°

This double star has a dramatic color contrast; the two stars have color indices of 3.6 and 0.0 respectively (Hipparcos). The brighter star varies irregularly in a 186-day period.

---

138 **δ Cephei** VARIABLE DOUBLE STAR IN CEPHEUS
SAO 34508 22^{h}29.2^m +58°25′
HIP 110991 Magnitudes 3.5–4.4, 6.1
NexStar Star 5267 Separation 40.9″
Position angle 191°

Prominent object. Delta Cephei, the prototype of the Cepheid variables, is also a double. The primary star varies from magnitude 3.5 to 4.4 in a very regular 5.37-day period.

The companion looks, to me, bluer than the primary. The position angle and separation are constant, so this is a good star on which to practice measuring double stars. A fine 8th-magnitude triple star is 6′ to the west.

---

139 **Krüger 60** DOUBLE RED DWARF STAR IN CEPHEUS
DO Cephei 22^{h}28.0^m +57°42′
GCVS 200137 Magnitudes 9.6, 11.2
HIP 110893 Separation 3.1″ (1999)
LX200 Star 348 Position angle 179° (1999)

Challenging object. Krüger 60 (Krueger 60, Kr 60) is one of the smallest stellar systems visible in amateur telescopes. This double star consists of two red dwarfs in a 45-year orbit the size of Saturn's orbit around the Sun (Figure 17.1).

The fainter component is a flare star (designated DO Cephei, GCVS 200137), which means that it sometimes brightens suddenly to equal or surpass the brighter one. The brighter component is also suspected of variability.

Krüger 60 is visible, but not easily split, in an 8-inch (20-cm) telescope. Do not use the finder chart in Burnham's *Celestial Handbook;* with a proper motion of nearly 1″ per year, the star has moved appreciably since Burnham's photograph was taken. Instead, rely on the accuracy of your pointing system, or star-hop from δ Cephei (the previous object in this list) using the chart in Figure 17.2.

Krüger 60 (WDS)

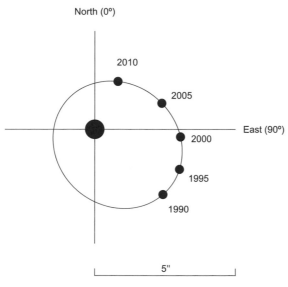

Figure 17.1. The orbit of Krüger 60, the smallest and fastest-orbiting stellar system visible in amateur telescopes, as seen through a telescope with a diagonal.

Currently, Krüger 60 is the northernmost star in a 30°–60°–90° triangle about 8′ long and is perceptibly reddish. A wide double star is 10′ to the northwest.

---

140 *η* **Cassiopeiae**   DOUBLE STAR IN CASSIOPEIA
SAO 21732   0^{h}49.1^m +57°49′
HIP 3821   Magnitudes 3.5, 7.3
NexStar Star 166   Separation 12.7″
                                              Position angle 318°

Prominent object. This is a fine double, cream-colored and reddish or purple. The secondary is a red dwarf. There are two additional, fainter companions on the other side of the primary.

---

141 **Pac-Man Nebula**   OPEN CLUSTER WITH NEBULA IN CASSIOPEIA
NGC 281   0^{h}52.8^m +56°36′
                                    Magnitude 7
                                    Diameter 30′

This bright star cluster is accompanied by faint nebulosity whose shape, in photographs, resembles the notorious video-game character of the 1980s.

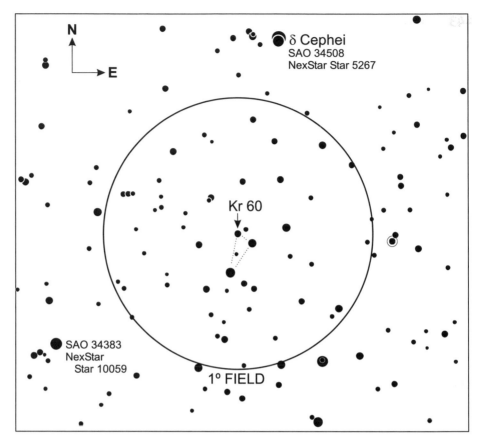

Figure 17.2. Identification chart for Krüger 60, matching the view in a telescope with a diagonal. Adapted from a chart generated with *TheSky* software, copyright 2001 Software Bisque, Inc., used by permission.

In the center of the cluster is the close quadruple star Burnham 1 (β 1, mags. 8.6, 9.3, 8.9, 9.7, seps. 1.4″, 3.8″, 9.0″).

---

142 Schedar
α Cassiopeiae
SAO 21609
HIP 3179
LX200 Star 7
NexStar Star 126

DOUBLE STAR IN CASSIOPEIA
0ʰ40.5ᵐ +56°32′
Magnitudes 2.4, 9.0
Separation 65″
Position angle 282°

A wide double star discovered in 1781 by Sir William Herschel.

---

143 NGC 7394

ASTERISM OR OPEN CLUSTER IN LACERTA
22^{h}50.6^m +52°10′
Magnitude 8
Diameter 15′

Listed as "nonexistent" in Sulentic and Tifft's 1973 revision of the *New General Catalogue,* and as "coordinates only" in the built-in catalogue of the LX200, this is actually a fairly striking chain or double chain of 10th- and 11th-magnitude stars with one considerably brighter star, visible in the 8-inch (20-cm) at 50×, more striking at 140×.

144 **Blue Snowball**
NGC 7662
Caldwell 22

PLANETARY NEBULA IN ANDROMEDA
23^{h}25.9^m +42°33′
Magnitude 8.6
Diameter 15″

Fairly prominent object. A planetary nebula with relatively high surface brightness and large size; a fine sight in the 8-inch (20-cm) even at 63×.

145 **Andromeda Galaxy**
M31
NGC 224

SPIRAL GALAXY IN ANDROMEDA
0^{h}42.7^m +41°16′
Magnitude 4
Size 1°×4°

Prominent object. This is the finest external galaxy in the sky; it is 3 million light-years away and is easily visible to the naked eye. Satellite galaxies M32 and M110 are in the same low-power field.

The center of M31 is very bright and has a starlike nucleus. A dust lane just south of the nucleus can be seen under a clear, dark sky.

The satellite galaxy M32 is bright, round, and easily mistaken for a star at low power. M110 is more diffuse and requires a dark sky.

146 **8 Lacertae**
SAO 72509
HIP 111546

ASTERISM IN LACERTA
22^{h}35.9^m +39°38′
Magnitude 5.6

This is a striking multiple star with at least four components in an arc-shaped arrangement 80″ long. Hipparcos data indicate that none of the stars are physically coupled.

147 NGC 752

OPEN CLUSTER IN ANDROMEDA
1^{h}57.8^m +37°41′
Magnitude 6
Diameter 1°

This rich cluster of faint stars more than fills the field at 50×. At one edge is the bright, wide double star 56 Andromedae (SAO 55107, NexStar Star 357, mags. 5.8 and 6.0, sep. 201″, position angle 298°).

---

148 **Hidden Galaxy**
NGC 404

ELLIPTICAL GALAXY IN ANDROMEDA
1^{h}09.4^m +35°43′
Magnitude 10
Diameter 2′

This relatively bright galaxy is omitted from many star atlases because its symbol would overlap with the bright reddish star β Andromedae. I had no trouble seeing it in an 8-inch (20-cm) telescope at 60× even under fairly unfavorable conditions. It looks like a fuzzy star just 6′ from β Andromedae.

---

149 **Caldwell 30**
NGC 7331

SPIRAL GALAXY IN PEGASUS
22^{h}37.1^m +34°25′
Magnitude 9.5
Size 11′×4′

Relatively bright as galaxies go, but not in Messier's list. This is a spiral seen at approximately the same angle as M31, with a bright core. It is fainter than M32 but brighter than M110.

---

150 $\pi^{1,2}$ **Pegasi**
SAO 72064, 72077
HIP 109352, 109410
NexStar Star 5186, 5189

WIDE DOUBLE STAR IN PEGASUS
22^{h}09.2^m +33°10′
Magnitudes 5.7, 4.3
Separation 573″ (=9.5′)
Position angle 88°

Prominent object. Pi-1 and Pi-2 Pegasi (also known as 27 and 29 Pegasi) constitute an unusually wide but genuine double star about 260 light-years away; the stars are about 0.7 light-year apart. Pi-1 also has a 10th-magnitude companion at 70″. Other stars in the field may be part of the system. Use low power.

---

151   **Pinwheel Galaxy**   SPIRAL GALAXY IN TRIANGULUM
          M33   1^{h}33.9^m +30°39′
          NGC 598   Magnitude 5.7
                                                 Size 70′×45′

Somewhat challenging object. Despite its high total brightness, this galaxy is spread
out over a patch of sky larger than the full moon, so it is easy to miss. To see it,
use low power and look for a "stain" of faint gray on the sky background. The
core, about 10′ in diameter, is brighter than the periphery.

Under clear, dark skies, substantial structure is visible; one hydrogen-rich
region in M33 has a separate NGC number (NGC 604, next item) and the sky all
around the galaxy looks lumpy. O'Meara reports seeing the spiral arms of M33
in a 4-inch telescope. Like M31, M33 is about 2 million light-years away.

---

152   **NGC 604**   NEBULA IN EXTERNAL GALAXY M33
                                                 1^{h}34^{m}30 +30°48′
                                                 Magnitude 10
                                                 Diameter 30″

Somewhat challenging object. This neglected but not unduly difficult nebula may
be your only opportunity to observe an object in an external galaxy – not the
galaxy as a whole, but one specific object within it. (But see also M54, p. 219.)

NGC 604 is a nebula much larger than the Orion Nebula, but located in the
galaxy M33, two million light-years away. From Earth, it is often easier to see
than M33 itself because of its higher surface brightness. It is plainly visible in
an 8-inch (20-cm) telescope under clear, dark skies and has reportedly been
glimpsed in a 2.4-inch (6-cm).

---

153   **36 Andromedae**   DOUBLE STAR IN ANDROMEDA
          SAO 74359   0^{h}55.0^m +23°38′
          HIP 4288   Magnitudes 6.1, 6.5
          LX200 Star 256   Separation 0.9″
          NexStar Star 181   Position angle 309

This close double star is a test for 6- to 8-inch (15- to 20-cm) telescopes; at 0.9″, it
is right at the Rayleigh limit for a 6-inch (15-cm). Use high power in steady air.

Uranus (LX200 Star 907) is in this area through 2020. At magnitude 6 and angular
diameter 3″, it is relatively easy to distinguish from a star, but look carefully. At

dark-sky sites, Uranus is just within the reach of naked-eye visibility; in even a small telescope, its blue-green color makes it stand out.

---

154 ψ^1 **Piscium**
SAO 74482
HIP 5131
NexStar Star 209

DOUBLE STAR IN PISCES
$1^h05.7^m +21°28'$
Magnitudes 5.3, 5.4
Separation 30″
Position angle 160°

Prominent object. Psi-1 Piscium is a fine wide double star for viewing at low or high power in even the smallest telescopes. It consists of two well-matched bright white stars.

---

155 **55 Piscium**
SAO 74182
HIP 3138
LX200 Star 252
NexStar Star 124

DOUBLE STAR IN PISCES
$0^h39.9^m +21°26'$
Magnitudes 5.6, 8.5
Separation 6.6″
Position angle 195°

A fine yellow and blue double star, similar to Albireo but smaller, requiring medium to high power.

---

156 **51 Pegasi**
SAO 90896
HIP 113357
NexStar Star 5365

STAR WITH PLANET IN PEGASUS
$22^h57.5^m +20°46'$
Magnitude 5.5

This ordinary-looking 5th-magnitude star was found, in 1995, to have a planet orbiting it, the first of a string of discoveries of extrasolar planets. This planet is half as heavy as Jupiter but orbits very close to the star, only one twentieth as far as the Earth is from the Sun, with an orbital period of 4 days. Its existence was inferred from Doppler shifts in the star's light; the planet itself cannot be seen.

---

157 γ **Arietis**
SAO 92680
HIP 8832
LX200 Star 262
NexStar Star 345

DOUBLE STAR IN ARIES
$1^h53.5^m +19°18'$
Magnitudes 4.5, 4.6
Separation 7.4″
Position angle 0°

Prominent object. Gamma Arietis, also known as **Mesarthim,** is a handsome double star with well-matched components. The stars are aligned exactly north–south. Use medium power.

---

158 | **19 Piscium** | CARBON STAR IN PISCES
TX Piscium
GCVS 660031
SAO 128374
HIP 117245

23ʰ46.4ᵐ +3°29′
Magnitude 4.5–5.3

The brightest carbon star in the sky, with color index 2.5 (reddish but not among the reddest stars). Its brightness varies irregularly. Because it is so red, catalogues based on blue-sensitive photography list it as much fainter.

---

159 | **CY Aquarii** | SHORT-PERIOD VARIABLE STAR IN AQUARIUS
GCVS 040125
HIP 111719

22ʰ37.8ᵐ +1°32′
Magnitude 10.4–11.2
Period 88 minutes

This extremely short-period variable star takes less than 10 minutes to rise from magnitude 11.6 to 10.4. Its light curve consists of a fast rise, a narrow peak, and a slow decline. Compare CY Aquarii to the other stars whose magnitudes are given in Figure 17.3. You can easily observe a full cycle or two in a single evening.

The precise position of CY Aquarii is 22ʰ37ᵐ47ˢ +1°32′04″, but the best way to get to it is probably to go to η (Eta) Aquarii (SAO 146181, NexStar Star 5287) and star-hop using the chart.

---

160 | **ζ Aquarii** | DOUBLE STAR IN AQUARIUS
SAO 146107
HIP 110960
NexStar Star 5263

22ʰ28.8ᵐ −0°01′
Magnitudes 4.4, 4.5
Separation 2.0″
Position angle 185°

This close double star is a good test for a 3.5-inch (9-cm) telescope; use high power. The components have been designated ζ^1 and ζ^2 (Zeta-1 and Zeta-2).

Because this star is on the celestial equator, you can use it to measure the field of view of your telescope. With the clock drive turned off, it will move one arc-minute every four seconds of time. When timing its passage, make sure it passes through the exact center of the field.

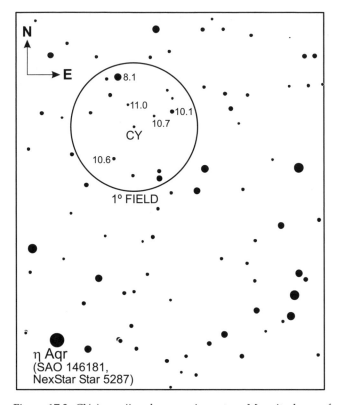

Figure 17.3. CY Aquarii and comparison stars. Magnitudes are from the Tycho catalogue. Adapted from a chart generated with *TheSky* software, copyright 2001 Software Bisque, Inc., used by permission.

161 41 Aquarii
SAO 190986
HIP 109786

DOUBLE STAR IN AQUARIUS
22^{h}14.3^m −21°04′
Magnitudes 5.6, 6.7
Separation 5.1″
Position angle 111°

A double star with a mild color contrast, cream-colored and bluish. There is also a 9th-magnitude companion much farther away, at 210″ and position angle 44°.

162 Sculptor Galaxy
NGC 253
Caldwell 65

GALAXY IN SCULPTOR
0^{h}47.6^m −25°17′
Magnitude 7.1
Size 25′ × 7′

A surprisingly bright galaxy in a barren area of the sky; it requires a relatively dark sky and low power. Look for an object resembling M110, but somewhat brighter and larger. Despite its high total brightness (magnitude 7.1), no part of it is as bright as M32.

This galaxy is physically similar to M31, but farther away, and is sometimes called the **Andromeda Galaxy of the South**. It is the finest galaxy south of the celestial equator – which goes to show you why the Australians envy our view of M31.

This is not to be confused with the **Sculptor System**, a nearby but extremely faint dwarf galaxy beyond the reach of amateur telescopes.

---

163 **R Sculptoris**   CARBON STAR IN SCULPTOR
GCVS 740001   1^{h}27.0^m −32°33′
SAO 193122   Magnitude 6.5–8.1
HIP 6759
NexStar Star 5815

This unusually red star (color index 5.0) varies semi-regularly between visual magnitudes 6.5 and 8.1 in a 370-day period. Because it is so red, many catalogues list it as much fainter (9th to 12th magnitude).

---

164 $\pi^{1,2}$ **Gruis**   DOUBLE STAR IN GRUS
SAO 231105, 231111   22^{h}22.7^m −45°57′
HIP 110478, 110506   Magnitudes 6.4, 5.6
NexStar Star 10057, 5243   Separation 258″
                                                              Position angle 254°

Prominent object. An unusually wide double with a color contrast, suitable for viewing at low power with even the smallest telescopes. Pi-1 (π^1) has a color index of 3 (quite reddish), while Pi-2 (π^2) is pure white. The two stars are different distances from Earth (500 and 130 light-years respectively) and are not physically connected.

The faint 11th-magnitude galaxy IC 5201 is in the same low-power field.

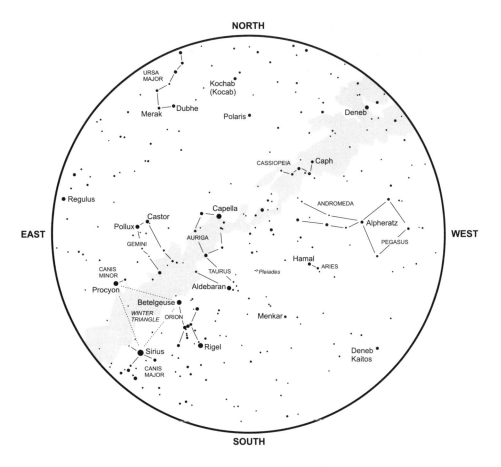

THE SKY AT SIDEREAL TIME 4:00

Mid-November	Midnight local time
Mid-December	10 p.m. local time
Mid-January	8 p.m. local time

As seen from latitude 40° north.

British and Canadian observers will see more stars above the north horizon;
southern U.S. observers will see more stars above the south horizon.
Based on a chart created with *Starry Night Pro* astronomy software
(http://www.starrynight.com), reproduced by permission.

Chapter 18
The November–December sky
(R.A. 2^h–6^h)

165 ι **Cassiopeiae**
SAO 12298
HIP 11569
LX200 Star 269
NexStar Star 463

TRIPLE STAR IN CASSIOPEIA
2^{h}29.1^m +67°24′
Magnitudes 4.6, 6.9, 9.0
Separations 2.8″, 7.1″
Position angles 231°, 116°

At 50×, Iota Cassiopeiae is a fine double star whose primary component looks a bit odd-shaped. Higher power reveals that this is actually a triple system.

Like many multiple stars, this one was discovered in 1829 by F. G. W. Struve, who designated it Σ 262. There is also an 8.5-magnitude star at 210″ in position angle 60°, apparently not a component of the system.

---

166 **U Camelopardalis**
GCVS 110004
SAO 12870
HIP 17257

CARBON STAR IN CAMELOPARDALIS
3^{h}41.8^m +62°39′
Magnitude 8.1–8.6

This low-amplitude variable is the reddest star in *Sky Catalogue 2000.0*, which gives its color index as 4.29; Hipparcos gives 4.91. However, V Aquilae may be redder (see p. 216).

If U Camelopardalis is not in your telescope's built-in catalogue, find it by the coordinates and look for a bright red star, or go to SAO 12874 (NexStar Star 713) and slew half a degree to the south (away from Polaris).

Double Cluster   PAIR OF OPEN CLUSTERS IN PERSEUS

167 **NGC 869**   2^{h}19.0^m +57°09′

h Persei   Magnitude 4.3

Caldwell 14 (A)   Diameter 30′

168 **NGC 884**   2^{h}22.4^m +57°07′

χ Persei   Magnitude 4.4

Caldwell 14 (B)   Diameter 30′

Prominent objects. These are perhaps the finest star clusters that Messier left out of his list. They are clearly visible to the naked eye even in suburban skies. They were designated h and χ (Chi) Persei respectively by Johannes Bayer (1604), who treated them as stars.

The two clusters are close together in space and are part of the Perseus OB1 Association, a large group of stars that were formed together. Three variable stars in NGC 884 are charted in *Sky & Telescope*, December 1994, p. 47.

---

169 **M34**   OPEN CLUSTER IN PERSEUS

NGC 1039   2^{h}42.1^m +42°45′

                                              Magnitude 5.2

                                              Diameter 40′

Prominent object. "A first-rate cluster for small telescopes" (O'Meara); also a fine sight in binoculars. Because many of the cluster stars are quite bright, M34 is impressive even under poor conditions. Several double stars are prominent.

---

170 *γ* **Andromedae**   DOUBLE STAR IN ANDROMEDA

SAO 37734   2^{h}03.9^m +42°21′

HIP 9640   Magnitudes 2.3, 5.0

NexStar Star 389   Separation 9.8″

                                              Position angle 64°

Prominent object. Also known as **Almach**, Gamma Andromedae is a handsome double star with a color contrast, yellow and blue-white. The fainter component is itself a very close double (mags. 5 and 6, separation 0.5″, gradually widening), resolvable in larger amateur telescopes.

Auriga Clusters   THREE OPEN CLUSTERS IN AURIGA

171 **M36**   5^{h}36.3^m +34°08′

NGC 1960   Magnitude 6.0

                                              Diameter 10′

172 **M37** 5^h52.3^m +32°33'
NGC 2099 Magnitude 5.6
Diameter 15'

173 **M38** 5^h28.7^m +35°51'
NGC 1912 Magnitude 6.4
Diameter 15'

These three clusters are all about 4000 light-years from us; they form a more spread-out system resembling the Double Cluster in Perseus. All are fine sights even in relatively small telescopes.

M36 is dominated by a zigzag stream of stars, with fainter stars clumped around it. M37 is more uniform but has a system of dark lanes (starless regions or perhaps dark nebulosity). M38 has a very prominent dark lane running right through the center.

---

174 **NGC 1514** PLANETARY NEBULA IN TAURUS
4^h09.2^m +30°47'
Magnitude 9
Diameter 1.5'

This planetary nebula is a relatively large patch of gas surrounding a 9th-magnitude star, midway between two 8th-magnitude stars 15' apart. It was discovered in 1790 by Sir William Herschel who described it as "a star of about 8th magnitude with a faint luminous atmosphere." This marked a turning point in the study of nebulae; previously they were all thought to be unresolved star clusters.

---

175 **Pleiades** OPEN CLUSTER IN TAURUS
Seven Sisters 3^h46.3^m +23°57'
M45 Magnitude 2
Diameter 1.5°

Extremely prominent object. The Pleiades or Seven Sisters may be the most impressive star cluster in the sky. The cluster has a vaguely dipper-like shape and I have encountered people who call it the Little Dipper (a name more traditionally ascribed to Ursa Minor, which looks like a dipper on maps but not in real life).

We commonly think of the cluster as containing six or seven stars visible to the naked eye. But the ancient Greeks named nine of the stars after Pleione, Atlas, and their seven daughters. Michael Maestlin, Kepler's teacher, reportedly saw 14 Pleiades and mapped 11 before the invention of the telescope.

---

176 *β* 536
South 437
SAO 76169, 76167

TRIPLE STAR IN PLEIADES CLUSTER
3^h46.3^m +24°11′
Magnitudes 8.1, 9.4, 7.7
Separations 0.9″, 39″
Position angles 182°, 308°

Right in the middle of the Pleiades cluster is a fine wide 8th-magnitude double star first catalogued by James South in 1823. It is an attractive pair that you can hardly miss while scanning the Pleiades.

In 1878, S. W. Burnham (abbreviated *β*) reported that one component of this pair was itself a close double; the separation was then 0.4″.

After Burnham discovered it, the close pair narrowed and widened again (Figure 18.1). Today the separation is about 0.9″ and is within easy reach of a 10-inch (25-cm) telescope at high power in steady air, but in a few years it will narrow again. In 2400, it will be a wide, easy telescopic double.

Catalogues do not entirely agree as to *which* element of the wide pair is the close double. In any triple system, the stars are designated A, B, and C, from

β 536 AB (WDS)

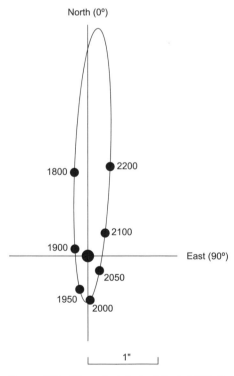

Figure 18.1. The calculated orbit of *β* 536, the close double star in the middle of the Pleiades, from data published in WDS. Since only part of an orbit has been observed, this is somewhat uncertain. The diagram matches the view through a telescope with a diagonal.

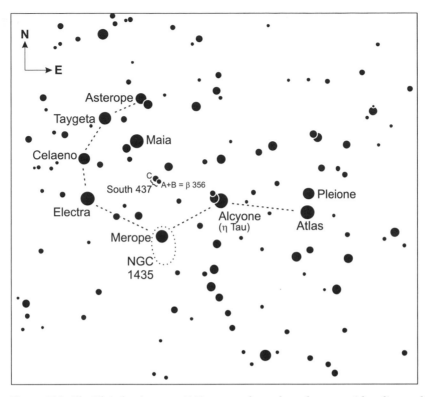

Figure 18.2. The Pleiades (to mag. 11.5) as seen through a telescope with a diagonal. For pronunciation of star names see p. 87. Adapted from a chart generated with *TheSky* software, copyright 2001 Software Bisque, Inc., used by permission.

the brightest star outward. Stars A and B are the close pair, and C is the more distant companion.

But in this case C is actually brighter than A, presumably because South was not good at making subtle magnitude distinctions; hence A and C are easy to mix up. Some published measurements of A and C give the position angle backward (off by 180°). I think I've sorted it all out in Figure 18.2, but have a look to make sure.

---

177 Merope Nebula
Tempel's Nebula
NGC 1435

REFLECTION NEBULA IN PLEIADES CLUSTER
3^{h}46.1^m +23°47′
Magnitude 10?
Diameter 10′

The Pleiades are embedded in a reflection nebula, IC 349. The brightest part of this is just south of Merope and is designated NGC 1435.

NGC 1435 is clearly visible at low power under dark skies; in fact at one point I realized I had been seeing it regularly and mistaking it for light scattered within the telescope. To distinguish the nebula from scattered light, compare Merope to other stars on the periphery of the cluster, and to stars in the Hyades cluster, which is free of nebulosity.

---

178 Crab Nebula
M1
NGC 1952

SUPERNOVA REMNANT IN TAURUS
5ʰ34.5ᵐ +22°01′
Magnitude 8.4
Size 6′×4′

Somewhat challenging object. Though it was the object with which Messier began his famous catalogue, the Crab Nebula is not particularly bright when seen in an amateur-sized telescope from suburban skies. Look carefully for a dim, irregular glow about 6′ wide, elongated, somewhat like a crab shell. Because of its unusual spectrum, narrow-band nebula filters do not do it justice; use a broadband filter.

M1 is what remains of a supernova that flared up in 1054 (as seen from Earth). In Messier's time it was three quarters of its present age and must have been more compact and somewhat brighter. It was discovered by John Bevis in 1731. At its center is a 16th-magnitude pulsar.

---

179 ε Arietis
SAO 75673
HIP 13914
NexStar Star 570

DOUBLE STAR IN ARIES
2ʰ59.2ᵐ +21°20′
Magnitudes 5.2, 5.6
Separation 1.4″
Position angle 209°

Epsilon Arietis is a close double that requires high power (150×) and steady air. It is a good test for a 3.5-inch (9-cm) telescope and is a rather good test of optical quality and atmospheric steadiness in somewhat larger telescopes.

---

180 NGC 1647

OPEN CLUSTER IN TAURUS
4ʰ46.0ᵐ +19°04′
Magnitude 6
Diameter 40′

A neglected open cluster that fills the low-power field. There is a wide 8th-magnitude double star at the northern edge.

---

181 $\theta^{1,2}$ **Tauri** DOUBLE STAR IN HYADES STAR CLUSTER
SAO 93955, 93957 4^{h}28.6^m +15°58′
HIP 20885, 20894 Magnitudes 3.4, 3.9
NexStar Star 926, 929 Separation 337″
Position angle 347°

Theta-1 and Theta-2 Tauri are a wide double star within the Hyades star cluster (Caldwell 41, Melotte 25). The cluster itself is 5° in diameter and does not fit in the telescopic field; it is the face of Taurus the Bull as seen with the naked eye.

---

182 **M78** REFLECTION NEBULA IN ORION
NGC 2068 5^{h}46.7^m +0°03′
Magnitude 8
Diameter 8′

This is the *other* Orion Nebula, a neglected object that looks very much like a comet. It is actually a pair of young, hot stars embedded in gas. The gas shines by reflected light, not fluorescence, so it photographs white, not red. It is the brightest reflection nebula in the sky.

A quarter of a degree to the north–northeast is a similar but fainter nebula, NGC 2071, surrounding a 10th-magnitude star.

---

183 **M77** SEYFERT GALAXY IN CETUS
NGC 1068 2^{h}42.7^m −0°01′
Magnitude 9
Diameter 7′

Prominent object. This galaxy has a *very* bright core and is accompanied by a star of comparable brightness just 90″ away. I have seen M77 in an 8-inch (20-cm) telescope at 140× even under a slightly hazy sky with a full moon. It is a good choice if you want to show someone a galaxy and the Moon or city lights are in the way.

The core of M77 varies irregularly from magnitude 10.4 to 11.2. The reason the core is so bright is that M77 is a Seyfert galaxy with a bright nucleus that emits radio waves. Seyfert galaxies, in turn, are smaller versions of quasars, the mysterious, powerful energy sources at the periphery of the known universe.

---

184 σ **Orionis**
SAO 132406
HIP 26549
NexStar Star 1259

MULTIPLE STAR IN ORION
5^{h}38.7^m −2°36′
Magnitudes 3.8, 6.6, etc.
Size 8′

Prominent object. The brightest star is a very close double (0.2″, not resolvable in amateur telescopes). It has companions at 11″, 13″, and 42″, in position angles 238°, 84°, and 62° respectively; the last of these is itself double, with a companion at 30″, position angle 232°.

All of this is merely the middle part of an arrow-shaped asterism 8′ long. The tail of the arrow is itself another multiple star.

---

185 **Mira**
o Ceti
SAO 129825
HIP 10826
NexStar Star 436

LONG-PERIOD VARIABLE STAR IN CETUS
2^{h}19.3^m −2°59′
Magnitude 3.4–9.3
Period 332 days

Now you see it, now you don't – at least, that was David Fabricius' experience when he discovered Mira in 1596. The name *Mira* means "remarkable," and this star, also designated Omicron Ceti, was the first pulsating variable to be recognized. Like other long-period variables, it is reddish in color and has a large amplitude.

For the next several years, Mira will be near minimum when high in the evening sky. Its period is a year minus a month. For the light curve see p. 133. Maxima are expected in late July 2002, June 2003, May 2004, April 2005, and March 2006. Even near minimum, Mira is quite red. There is a 9.2-magnitude star 2′ to the east.

---

186 **Orion Nebula**
θ^{1,2} Orionis
M42

EMISSION NEBULA IN ORION
5^{h}35.3^m −5°23′
Magnitude 3
Size (central portion) 1°

Extremely prominent object. M42 is the finest deep-sky object visible from the United States and Europe, and a lifetime can be spent contemplating its intricate structure.

The two brightest stars in the nebula are Theta-1 and Theta-2 Orionis; they form a wide double. Theta-1, in turn, is the brightest element of the **Trapezium**, a roughly square quadruple-star system (Figure 18.3).

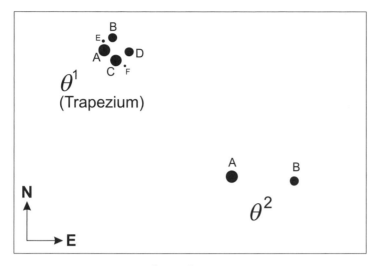

Figure 18.3. Multiple stars θ^1 and θ^2 Orionis at the center of the Orion Nebula, as seen through a telescope with a diagonal.

Many observers can see a fifth star (E) accompanying the Trapezium; some can see a sixth (F). The system is actually more than sextuple, since the double star θ^2 Orionis is also involved, as are other stars in the nebula.

Don't miss M43, the northern portion of the Orion Nebula. It surrounds another group of bright stars about half a degree to the north of the Trapezium.

---

187 **NGC 1999** EMISSION NEBULA IN ORION
5ʰ36.4ᵐ −6°43′
Magnitude 9
Diameter 15′

This compact cloud of gas surrounds a star of magnitude 10.3. I find it brighter than M78, though smaller. If it were not so close to M42, it would be a well-known showpiece.

---

188 o^2 **Eridani** TRIPLE STAR WITH WHITE DWARF IN ERIDANUS
40 Eridani 4ʰ15.3ᵐ −7°39′
SAO 131063 Magnitudes 4.4, 9.5, 11.2
HIP 19849 Separations 82″, 8.9″
NexStar Star 852 Position angles 104°, 337°

Omicron-2 Eridani is a triple star whose B component, 9th magnitude, is one of the few white dwarfs within reach of amateur telescopes. The dwarf's diameter is only 20 000 km, less than twice that of the Earth.

The white dwarf, in turn, has an 11th-magnitude red dwarf companion in a relatively rapid 248-year orbit.

---

189   Rigel   DOUBLE STAR IN ORION
β Orionis   5ʰ14.5ᵐ −8°12′
SAO 131907   Magnitudes 0.3, 10.4
HIP 24436   Separation 9.6″
LX200 Star 41   Position angle 202°
NexStar Star 1118

Prominent object. Rigel, or Beta Orionis, is the seventh brightest star in the sky. Despite its brightness, the companion is clearly visible at medium power.

---

190   ε Eridani   SUN-LIKE STAR WITH PLANET IN ERIDANUS
SAO 130564   3ʰ32.9ᵐ −9°28′
NexStar Star 694   Magnitude 3.7

This is what the Sun would look like from 10 light-years away. If there is life elsewhere in our galaxy, it's very likely to be here. Epsilon Eridani has a Jupiter-sized planet orbiting it in a Jupiter-like orbit with a period of 7 years. (The planet is not visible from Earth, of course; it is detected by its effect on the star's motion.) Epsilon Eridani may also have other planets in an arrangement similar to that of our Solar System.

---

191   IC 418   PLANETARY NEBULA IN LEPUS
                                                   5ʰ27.5ᵐ −12°42′
                                                   Magnitude 9
                                                   Diameter 12″

A tiny but bright planetary nebula that looks like a blue-green star at low power. Medium power brings out its true nature.

---

192   NGC 1535   PLANETARY NEBULA IN ERIDANUS
                                                   4ʰ14.3ᵐ −12°44′
                                                   Magnitude 10
                                                   Diameter 20″

A planetary nebula with a bright central disk and an outer halo, described by Dreyer as "mottled."

---

193   **Hind's Crimson Star**   LONG-PERIOD VARIABLE STAR IN LEPUS
R Leporis   4^{h}59.6^m −14°48′
GCVS 480001   Magnitude 5.5–11.7
SAO 150058   Period 427 days
HIP 23203
NexStar Star 1054

This Mira-type variable was discovered by J. R. Hind in 1845. When near minimum, it looks insignificant even in an 8-inch (20-cm) telescope. Because the period is considerably more than a year, you get to see it at a different point in each cycle each winter. Maxima are expected in September 2002, November 2003, January 2005, and March 2006.

The color index of R Leporis is given variously as 1.8 (older catalogues), 4.2 (Hipparcos), or 5.5 (Skiff). If this last value is correct, it is indeed one of the reddest stars in the sky.

---

194   **Perfect Right Angle**   ASTERISM IN CETUS
v Ceti   2^{h}00.0^m −21°05′
SAO 167471   Magnitude 4.0
NexStar Star 372

Nearby 5th- and 7th-magnitude stars form a right triangle with Upsilon Ceti at the apex.

The right angle is nearly perfect (89.9° according to Hipparcos data). The companions are each almost exactly $\frac{1}{4}$° away. Indeed, Upsilon Ceti is a rich source of exact numbers – its right ascension is exactly 2 hours (epoch 2000.0) and its magnitude is exactly 4.0.

---

195   **M79**   GLOBULAR CLUSTER IN LEPUS
NGC 1904   5^{h}24.2^m −24°31′
                                                     Magnitude 7.7
                                                     Diameter 6′

"A smattering of fairly bright members against a blazing core" (O'Meara). Relatively low surface brightness for a globular.

---

196   **h 3752**   TRIPLE STAR IN LEPUS
        SAO 170352 $5^h21.8^m$ $-24°46'$
        HIP 25045   Magnitudes 5.4, 6.6, 9.2
        NexStar Star 6506   Separations 3.5″, 62″
                                                    Position angles 93°, 105°

This triple star is less than a degree west of M79. It was discovered by Sir John
Herschel in 1837. The primary is yellow-white, the secondary is blue-white, and
the more distant 9th-magnitude companion is reddish.

---

197   **NGC 1360**   PLANETARY NEBULA IN FORNAX
                                                        $3^h33.3^m$ $-25°51'$
                                                                Magnitude 9
                                                                Diameter 7′

A relatively large but neglected planetary nebula with a visible central star.
Surface brightness is relatively low.

       Fornax Galaxy Cluster   CLUSTER OF GALAXIES IN FORNAX
198   **NGC 1399** $3^h38.5^m$ $-35°27'$
                                                            Magnitude 8.8
                                                            Size 7′×8′

199   **NGC 1404** $3^h38.9^m$ $-35°35'$
                                                            Magnitude 9.7
                                                            Size 5′×4′

These two elliptical galaxies are the brightest members of a very rich cluster
located about 50 million light-years from us. You may be able to see as many as
ten galaxies brighter than magnitude 12 within a 1° field.

---

200   **Caldwell 73**   GLOBULAR CLUSTER IN COLUMBA
        NGC 1851 $5^h14.1^m$ $-40°03'$
                                                            Magnitude 7.3
                                                            Diameter 11′

The *New General Catalogue* describes this neglected globular cluster as "!, vB, vL,
R, vsvvbM, rrr," which means, "Remarkable, very bright, very large, rich, very
suddenly much brighter in the middle, well resolved into stars." Like several
other globulars, it contains an X-ray source.

Columba the Dove is the only constellation with Biblical rather than classical roots. It was apparently created in the 1500s to represent the dove returning to the ship Argo, reinterpreted as Noah's Ark. Those of us who do not find religious awe incompatible with astronomy may wish to pause, look up at the splendors of the late autumn sky, and ponder for a moment the words of the prophet Amos (5:8):

He who made the Pleiades and Orion ... THE LORD *is His name.*

Appendix A
Converting decimal minutes to seconds

Some catalogues use decimal minutes of right ascension and declination; others use minutes and seconds. To interconvert the two, use the chart, or simply multiply the first decimal digit by 6 and treat it as seconds.

$0.0^m = 00^s$	$0.5^m = 30^s$
$0.1^m = 06^s$	$0.6^m = 36^s$
$0.2^m = 12^s$	$0.7^m = 42^s$
$0.3^m = 18^s$	$0.8^m = 48^s$
$0.4^m = 24^s$	$0.9^m = 54^s$

Recall that a minute of right ascension is 15 times as big as a minute of declination. Normally, seconds of declination can be ignored.

Appendix B
Precession from 1950 to 2000

Many older maps still use epoch 1950 coordinates. To convert right ascensions and declinations from 1950 to 2000, find the entry in Table B.1 nearest the object's position in the sky and add the corrections indicated there.

To convert 2000 coordinates to 1900 for AAVSO variable-star designations and the like, double the tabulated correction and apply it in the opposite direction (subtracting instead of adding and vice versa).

To calculate precession, let α and δ stand for the object's initial right ascension and declination respectively, both expressed in degrees. (Recall that $1^h = 15°$.) Let N stand for the number of years between epochs (negative if you are converting from a later epoch to an earlier one). Then compute:

R.A. correction (in seconds) $= (3.073 + 1.336 \sin\alpha \tan\delta) \times N$

Declination correction (in arc-seconds) $= (20.042 \cos\alpha) \times N$

and add the corrections to the original R.A. and declination.

These formulae are for dates within a couple of centuries of 2000 and positions at least $0.1°$ away from the celestial poles. For more accurate formulae see the *Astronomical Almanac* published by the U.S. and British governments.

Table B.1. *Precession for 50 years*

Dec.	0ʰ	3ʰ	6ʰ	9ʰ	R.A. 12ʰ	15ʰ	18ʰ	21ʰ	24ʰ
+88°	RA +2.6^m Dec +17′	RA +25.1^m Dec +12′	RA +34.4^m Dec 0′	RA +25.1^m Dec −12′	RA +2.6^m Dec −17′	RA −20.0^m Dec −12′	RA −29.3^m Dec 0′	RA −20.0^m Dec +12′	RA +2.6^m Dec +17′
+85°	RA +2.6^m Dec +17′	RA +11.6^m Dec +12′	RA +15.3^m Dec 0′	RA +11.6^m Dec −12′	RA +2.6^m Dec −17′	RA −6.4^m Dec −12′	RA −10.2^m Dec 0′	RA −6.4^m Dec +12′	RA +2.6^m Dec +17′
+80°	RA +2.6^m Dec +17′	RA +7.0^m Dec +12′	RA +8.9^m Dec 0′	RA +7.0^m Dec −12′	RA +2.6^m Dec −17′	RA −1.9^m Dec −12′	RA −3.8^m Dec 0′	RA −1.9^m Dec +12′	RA +2.6^m Dec +17′
+70°	RA +2.6^m Dec +17′	RA +4.7^m Dec +12′	RA +5.6^m Dec 0′	RA +4.7^m Dec −12′	RA +2.6^m Dec −17′	RA +0.4^m Dec −12′	RA −0.5^m Dec 0′	RA +0.4^m Dec +12′	RA +2.6^m Dec +17′
+60°	RA +2.6^m Dec +17′	RA +3.9^m Dec +12′	RA +4.5^m Dec 0′	RA +3.9^m Dec −12′	RA +2.6^m Dec −17′	RA +1.2^m Dec −12′	RA +0.6^m Dec 0′	RA +1.2^m Dec +12′	RA +2.6^m Dec +17′
+50°	RA +2.6^m Dec +17′	RA +3.5^m Dec +12′	RA +3.9^m Dec 0′	RA +3.5^m Dec −12′	RA +2.6^m Dec −17′	RA +1.6^m Dec −12′	RA +1.2^m Dec 0′	RA +1.6^m Dec +12′	RA +2.6^m Dec +17′
+40°	RA +2.6^m Dec +17′	RA +3.2^m Dec +12′	RA +3.5^m Dec 0′	RA +3.2^m Dec −12′	RA +2.6^m Dec −17′	RA +1.9^m Dec −12′	RA +1.6^m Dec 0′	RA +1.9^m Dec +12′	RA +2.6^m Dec +17′
+30°	RA +2.6^m Dec +17′	RA +3.0^m Dec +12′	RA +3.2^m Dec 0′	RA +3.0^m Dec −12′	RA +2.6^m Dec −17′	RA +2.1^m Dec −12′	RA +1.9^m Dec 0′	RA +2.1^m Dec +12′	RA +2.6^m Dec +17′
+20°	RA +2.6^m Dec +17′	RA +2.8^m Dec +12′	RA +3.0^m Dec 0′	RA +2.8^m Dec −12′	RA +2.6^m Dec −17′	RA +2.3^m Dec −12′	RA +2.2^m Dec 0′	RA +2.3^m Dec +12′	RA +2.6^m Dec +17′
+10°	RA +2.6^m Dec +17′	RA +2.7^m Dec +12′	RA +2.8^m Dec 0′	RA +2.7^m Dec −12′	RA +2.6^m Dec −17′	RA +2.4^m Dec −12′	RA +2.4^m Dec 0′	RA +2.4^m Dec +12′	RA +2.6^m Dec +17′

Dec.									
0°	RA $+2.6^m$ Dec $+17'$	RA $+2.6^m$ Dec $+12'$	RA $+2.6^m$ Dec $0'$	RA $+2.6^m$ Dec $-12'$	RA $+2.6^m$ Dec $-17'$	RA $+2.6^m$ Dec $-12'$	RA $+2.6^m$ Dec $0'$	RA $+2.6^m$ Dec $+12'$	RA $+2.6^m$ Dec $+17'$
−10°	RA $+2.6^m$ Dec $+17'$	RA $+2.4^m$ Dec $+12'$	RA $+2.4^m$ Dec $0'$	RA $+2.4^m$ Dec $-12'$	RA $+2.6^m$ Dec $-17'$	RA $+2.7^m$ Dec $-12'$	RA $+2.8^m$ Dec $0'$	RA $+2.7^m$ Dec $+12'$	RA $+2.6^m$ Dec $+17'$
−20°	RA $+2.6^m$ Dec $+17'$	RA $+2.3^m$ Dec $+12'$	RA $+2.2^m$ Dec $0'$	RA $+2.3^m$ Dec $-12'$	RA $+2.6^m$ Dec $-17'$	RA $+2.8^m$ Dec $-12'$	RA $+3.0^m$ Dec $0'$	RA $+2.8^m$ Dec $+12'$	RA $+2.6^m$ Dec $+17'$
−30°	RA $+2.6^m$ Dec $+17'$	RA $+2.1^m$ Dec $+12'$	RA $+1.9^m$ Dec $0'$	RA $+2.1^m$ Dec $-12'$	RA $+2.6^m$ Dec $-17'$	RA $+3.0^m$ Dec $-12'$	RA $+3.2^m$ Dec $0'$	RA $+3.0^m$ Dec $+12'$	RA $+2.6^m$ Dec $+17'$
−40°	RA $+2.6^m$ Dec $+17'$	RA $+1.9^m$ Dec $+12'$	RA $+1.6^m$ Dec $0'$	RA $+1.9^m$ Dec $-12'$	RA $+2.6^m$ Dec $-17'$	RA $+3.2^m$ Dec $-12'$	RA $+3.5^m$ Dec $0'$	RA $+3.2^m$ Dec $+12'$	RA $+2.6^m$ Dec $+17'$
−50°	RA $+2.6^m$ Dec $+17'$	RA $+1.6^m$ Dec $+12'$	RA $+1.2^m$ Dec $0'$	RA $+1.6^m$ Dec $-12'$	RA $+2.6^m$ Dec $-17'$	RA $+3.5^m$ Dec $-12'$	RA $+3.9^m$ Dec $0'$	RA $+3.5^m$ Dec $+12'$	RA $+2.6^m$ Dec $+17'$
−60°	RA $+2.6^m$ Dec $+17'$	RA $+1.2^m$ Dec $+12'$	RA $+0.6^m$ Dec $0'$	RA $+1.2^m$ Dec $-12'$	RA $+2.6^m$ Dec $-17'$	RA $+3.9^m$ Dec $-12'$	RA $+4.5^m$ Dec $0'$	RA $+3.9^m$ Dec $+12'$	RA $+2.6^m$ Dec $+17'$
−70°	RA $+2.6^m$ Dec $+17'$	RA $+0.4^m$ Dec $+12'$	RA -0.5^m Dec $0'$	RA $+0.4^m$ Dec $-12'$	RA $+2.6^m$ Dec $-17'$	RA $+4.7^m$ Dec $-12'$	RA $+5.6^m$ Dec $0'$	RA $+4.7^m$ Dec $+12'$	RA $+2.6^m$ Dec $+17'$
−80°	RA $+2.6^m$ Dec $+17'$	RA -1.9^m Dec $+12'$	RA -3.8^m Dec $0'$	RA -1.9^m Dec $-12'$	RA $+2.6^m$ Dec $-17'$	RA $+7.0^m$ Dec $-12'$	RA $+8.9^m$ Dec $0'$	RA $+7.0^m$ Dec $+12'$	RA $+2.6^m$ Dec $+17'$
−85°	RA $+2.6^m$ Dec $+17'$	RA -6.4^m Dec $+12'$	RA -10.2^m Dec $0'$	RA -6.4^m Dec $-12'$	RA $+2.6^m$ Dec $-17'$	RA $+11.6^m$ Dec $-12'$	RA $+15.3^m$ Dec $0'$	RA $+11.6^m$ Dec $+12'$	RA $+2.6^m$ Dec $+17'$
−88°	RA $+2.6^m$ Dec $+17'$	RA -20.0^m Dec $+12'$	RA -29.3^m Dec $0'$	RA -20.0^m Dec $-12'$	RA $+2.6^m$ Dec $-17'$	RA $+25.1^m$ Dec $-12'$	RA $+34.4^m$ Dec $0'$	RA $+25.1^m$ Dec $+12'$	RA $+2.6^m$ Dec $+17'$

Appendix C
Julian date, 2001–2015

The **Julian date** (**JD**) is the number of days elapsed since noon UT on January 1 of 4713 B.C. It is the standard way of giving the date and time of variable star observations and is also used in astronomy for other purposes.

The Julian date system was introduced in 1582 by Joseph Justus Scaliger, who named it in honor of his father Julius Scaliger; it has nothing to do with the Julian calendar of Julius Caesar. The starting date in 4713 B.C. was chosen for easy conversion between several ancient calendars.

Note that the Julian day begins at noon, not midnight. Astronomers also use the **modified Julian date** (**MJD**), which is the JD minus 2 400 000.5. The MJD day begins at midnight.

AAVSO publications also use the JD minus 2 400 000, ignoring the 0.5. With long-period variables, this makes little difference.

Table C.1 gives the MJD for 0:00 UT on the "zeroth" day of every month (i.e., last day of the previous month) from 2001 to 2020, with instructions for computing the JD.

The AAVSO publishes a calendar giving the Julian date for each day (including an online version at http://www.aavso.org), and most if not all computer sky chart programs give the Julian date.

Table C.1. *MJD at 0:00 UT one day before the first of each month, 2001–2015*

Year	Jan	Feb	Mar	Apr	May	Jun	Jul	Aug	Sep	Oct	Nov	Dec
2001	51 909	51 940	51 968	51 999	52 029	52 060	52 090	52 121	52 152	52 182	52 213	52 243
2002	52 274	52 305	52 333	52 364	52 394	52 425	52 455	52 486	52 517	52 547	52 578	52 608
2003	52 639	52 670	52 698	52 729	52 759	52 790	52 820	52 851	52 882	52 912	52 943	52 973
2004	53 004	53 035	53 064	53 095	53 125	53 156	53 186	53 217	53 248	53 278	53 309	53 339
2005	53 370	53 401	53 429	53 460	53 490	53 521	53 551	53 582	53 613	53 643	53 674	53 704
2006	53 735	53 766	53 794	53 825	53 855	53 886	53 916	53 947	53 978	54 008	54 039	54 069
2007	54 100	54 131	54 159	54 190	54 220	54 251	54 281	54 312	54 343	54 373	54 404	54 434
2008	54 465	54 496	54 525	54 556	54 586	54 617	54 647	54 678	54 709	54 739	54 770	54 800
2009	54 831	54 862	54 890	54 921	54 951	54 982	55 012	55 043	55 074	55 104	55 135	55 165
2010	55 196	55 227	55 255	55 286	55 316	55 347	55 377	55 408	55 439	55 469	55 500	55 530
2011	55 561	55 592	55 620	55 651	55 681	55 712	55 742	55 773	55 804	55 834	55 865	55 895
2012	55 926	55 957	55 986	56 017	56 047	56 078	56 108	56 139	56 170	56 200	56 231	56 261
2013	56 292	56 323	56 351	56 382	56 412	56 443	56 473	56 504	56 535	56 565	56 596	56 626
2014	56 657	56 688	56 716	56 747	56 777	56 808	56 838	56 869	56 900	56 930	56 961	56 991
2015	57 022	57 053	57 081	57 112	57 142	57 173	57 203	57 234	57 265	57 295	57 326	57 356

Julian date (JD) $= 2\,400\,000.5 + $ MJD from table above $+$ day of month $+ \dfrac{\text{hours since 0h UT}}{24}$

Modified Julian date (MJD) $=$ same, omitting $2\,400\,000.5$

Example: 2007 July 10, 3:00 UT $= 2\,400\,000.5 + 54\,281 + 10 + \dfrac{3}{24} = 2\,454\,291.625$

253

Index

Page numbers in italics refer to figures or tables.

Index